P9-AGN-546

Praise for *Beyond Biocentrism*

"Lanza and Berman employ cutting edge science to rediscover ancient truths about life and death and reconceptualize our very notions of reality and consciousness. *Beyond Biocentrism* is an enlightening and fascinating journey that will forever alter your understanding of your own existence."

—Deepak Chopra

"Robert Lanza and Bob Berman present an audacious program to restore meaning to science—to provide explanations that go deeper than today's physical theories. *Beyond Biocentrism* is a joyride through the history of science and cutting-edge physics, all with a very serious purpose: to find the long-overlooked connection between the conscious self and the universe around us."

—Corey S. Powell, editor at large and
former editor-in-chief, *Discover* magazine

"This intriguing and provocative book will challenge some of what you know and push you into rethinking your view of science—all the while entertaining you with a fast-paced, exhilarating narrative journey."

—David J. Eicher, editor-in-chief, *Astronomy* magazine

In *Beyond Biocentrism*, stem cell pioneer Robert Lanza, writing with astronomer Bob Berman, presents a lucid tour de force of his thrilling but controversial theory that consciousness creates reality and the cosmos itself. Will machines ever achieve consciousness? Are plants aware? Is death an illusion? These are some of the big questions tackled in *Beyond Biocentrism*, which serves up a new, biology-based theory of everything that is as delightful to read as it is fascinating. Tremendously clear and lovely writing—a huge achievement.

—Pamela Weintraub, psychology and health editor, *Aeon* magazine,
and former executive editor of *Discover* magazine and
editor-in-chief of *OMNI* (internet/magazine)

"Lanza and Berman's latest statement of their theory of 'biocentrism' changes the way we think about age-old religious questions such as the origin of the universe and human immortality. Based on cutting edge work in physics and biology and explained with exceptional clarity, *Beyond Biocentrism* is must reading for anyone interested in science and religion."

—Ronald M Green, professor emeritus for the
study of ethics and human values, and former chairman
of the Department of Religion at Dartmouth College

"*Beyond Biocentrism* delves further into the role of the observer and consciousness. It offers a neurobiological point of view to help answer questions about the world around us. Lanza and Berman make the journey towards a better understanding of the role of consciousness and perception. I liked the book quite a lot! It was a fun read."

—Kwang-Soo Kim, professor of psychiatry and neuroscience, Harvard Medical School, and director, Molecular Neurobiology Laboratory, McLean Hospital

"Lanza and Berman take the reader on a remarkable journey, setting out to prove that there is more to life and existence than we have assumed. They present scientific evidence that makes us re-consider everything we've thought to be true about the nature of reality. *Beyond Biocentrism* is a fascinating and thought-provoking 'must-read' book that shows us a new way of looking at the universe and ourselves."

—Anthony Atala, W. H. Boyce Professor, chairman, and director of the Wake Forest Institute for Regenerative Medicine, Wake Forest University

"There are few intellectual endeavors more thrilling than contemplating the role of human consciousness in creating reality and the universe, and Lanza and Berman bring to life the quest for understanding how that can possibly be so. If you know just enough physics to wonder whether the moon is still there when no one is looking at it, and even if you've never thought about anything so seemingly preposterous, you'll have a great time reading *Beyond Biocentrism*."

—Sharon Begley, senior science writer at *Stat*, and former science editor and correspondent for *Newsweek*, the *Wall Street Journal*, and *Reuters*

"*Beyond Biocentrism* is a must read for anyone who has ever wondered where modern science (and the weirdness of relativity and quantum mechanics) is going. What does it all mean? Brilliant and insightful. Few books come along in our lives that change the way we see the world. *Beyond Biocentrism* is such a book."

—Ralph D Levinson, health sciences professor, UCLA

BEYOND
BIOCENTRISM

BEYOND BIOCENTRISM

Rethinking Time, Space, Consciousness, and the Illusion of Death

ROBERT LANZA, MD,

with Bob Berman

BenBella Books, Inc.
Dallas, TX

Copyright © 2016 by Robert Lanza, MD, and Bob Berman

All rights reserved. No part of this book may be used or reproduced in any manner whatsoever without written permission except in the case of brief quotations embodied in critical articles or reviews.

BenBella Books, Inc.
10440 N. Central Expy., Suite 800
Dallas, TX 75231
www.benbellabooks.com
Send feedback to feedback@benbellabooks.com

Printed in the United States of America
10 9 8 7 6 5 4 3 2

Library of Congress Cataloging-in-Publication Data is available upon request.
LCCN: 2015043363
ISBN-13: 978-1-942952-21-3 (print) / 978-1-942952-22-0 (electronic)

Editing by Heather Butterfield
Copyediting by James Fraleigh
Proofreading by Michael Fedison, Brittney Martinez, and Rachel Phares
Indexing by Jigsaw Information
Text design and composition by Aaron Edmiston
Cover design by Brad Foltz
Illustrations on pages 48, 52, 66, 69, 72, 101–103, 114, 122–123, 168 by Jacqueline Rogers
Illustration on page 149 by Wim R. Euverman
Image on 156: plant (John Sims), octopus (A. Pollock), mouse (George Shuklin), sparrow (W. Wright), tadpole (rainforest_harley on Flickr)
Printed by Lake Book Manufacturing

Distributed by Perseus Distribution
www.perseusdistribution.com

To place orders through Perseus Distribution:
Tel: (800) 343-4499
Fax: (800) 351-5073
E-mail: orderentry@perseusbooks.com

Significant discounts for bulk sales (minimum of 25 copies) are available.
Please contact Aida Herrera at aida@benbellabooks.com.

CONTENTS

INTRODUCTION

Why do you insist the universe is not a conscious intelligence,
when it gives birth to conscious intelligences?
—Cicero, c. 44 BCE

The deepest, most vexing issues have not changed much since
the beginning of civilization. People eight thousand years ago
worried about death. Those in ancient Babylonia shared with us
an obsession with the passage of time. Thinkers in every culture
have pondered Earth and the heavens and generally have seen
them as existing in a space-based matrix. The nature of life and
consciousness started to obsess us as soon as we came down from
the forest roof and grew brains large enough to be tormented.

Tackling these big-ticket items has properly become a focus
for science as well. Our first book, *Biocentrism*, offered a very different way of looking at the universe and reality itself. Because
this perspective is so unlike the descriptions we are accustomed
to, it takes some time and thought to comprehend. That's what
this book is about.

This way of thinking starts by recognizing that our existing model of reality is looking increasingly creaky in the face of recent scientific discoveries. Science tells us with some precision that over 95 percent of the universe is composed of dark matter and dark energy, but it must confess that it doesn't really know what dark matter is and knows even less about dark energy. Science points more and more toward an infinite universe but has no ability to explain what that means. Concepts such as time, space, and even causality are increasingly being demonstrated as meaningless.

All of science is based on information passing through our consciousness, but science doesn't have a clue what consciousness is. Studies have repeatedly established a clear link between subatomic states and observation by conscious observers, but science cannot explain this connection in any satisfactory way. Biologists describe the origin of life as a random occurrence in a dead universe, but have no real understanding of how life began or why the universe appears to have been exquisitely designed for its emergence.

This new worldview is completely based on science and is better supported by the scientific evidence than traditional explanations. It challenges us to fully accept the implications of the latest scientific findings in fields ranging from plant biology and cosmology to quantum entanglement and consciousness.

If we listen to what the science is telling us, it becomes ever more clear that life and consciousness are fundamental to any true understanding of the universe. This new perception of the nature of the universe is called *biocentrism*.

If you read *Biocentrism*, welcome back for a deeper and more thorough exploration into the subject, including chapters that solely involve key issues such as death, and important ancillary investigations into topics such as awareness in the botanical world, how we gain information, and whether machines can ever become conscious.

REALITY 101

1

It's enough for me to be sure that you and I
exist at this moment.
—Gabriel García Márquez, *One Hundred Years of Solitude* (1967)

S omewhere around the age of seven, most kids ask uncomfortable questions. Is there an end to the universe? How did I get here? Some children, perhaps after a pet hamster has passed away, also start to worry about death.

A few venture even more deeply. They know they've come into a world that seems complex and mysterious but can still occasionally recall the remnant of clarity and joy that was theirs during the first year of life. But as they progress through middle and then high school, and science teachers provide the standard explanation of the cosmos, they shrug that remnant off. The framework of existence has become either droningly academic or else a mere matter of philosophy. If they ponder it occasionally as an adult, their usual takeaway is that the entire cosmological worldview seems confused and unsatisfying.

The most widely accepted model of the universe depends on the part of the world and the time in history in which the questions were posed. A few centuries ago, Church and Scripture provided the framework for the Big Picture. By the 1930s, biblical explanations were no longer in vogue among the intelligentsia and were eventually replaced by the *cosmic egg* model—where everything began with a sudden explosive event—similar to what Edgar Allan Poe originally proposed in an 1848 essay.

In this model, the universe was presented as a kind of self-operating machine. It was composed of stupid stuff, meaning atoms of hydrogen and other elements that had no innate intelligence. Nor did any sort of external intelligence rule. Rather, unseen forces such as gravity and electromagnetism, acting according to the random laws of chance, produced everything we observe. Atoms slammed into others. Clouds of hydrogen contracted to form stars. Leftover globs of matter orbiting these newborn suns cooled into planets.

Billions of lifeless years passed with the cosmos set on "automatic," until on at least one planet, and possibly others, life began. How this happened remains mysterious to our science. After all, we can take the known proteins, minerals, water, and everything else that an animal body contains and whirl it in a blender till the cows come home and still not have life.

If life and its genesis remain a mystery, consciousness is an enigma squared. For it is one thing to reproduce and grow and whatever else we deem to be life's characteristics; *awareness* is quite another feature. They are not the same. Yeast and HIV are alive. But do they *perceive*? Do all living creatures experience some analogue of our own sense of rapture at the deep purples in a twilight sky?

The issue is more than academic. For nearly a full century, physicists have seen that the observer's consciousness affects the results of experiments. Yet this has been little more than shrugged off as enigmatic and bewildering.

As for how consciousness could arise in the first place, no one even has guesses. We cannot fathom how lumps of carbon, drops of water, or atoms of insensate hydrogen ever came together and acquired a sense of smell. The issue is apparently too baffling to raise at all. Merely to bring up the topic of the origin of perception is to brand oneself a kook. Although former *Encyclopedia Britannica* publisher Paul Hoffman called it "the deepest problem in all of science," it usually sounds too odd and foreign to be discussed in serious venues. Nonetheless, we will later return robustly to the issue of consciousness. For now, suffice that its genesis is shrouded in a mystery as absolute as any inventory of the landfills near the New Jersey Turnpike.

So our standard model of the universe consists of an interesting mixture of the living and the nonliving. Both are part and parcel of a universe that, cosmology explains, burst out of nothingness 13.8 billion years ago, and the whole shebang keeps getting larger.

This is the story. Everyone has heard it. It's recited to school students throughout the world. And yet everyone can feel how vacuous and unsatisfying this narrative is.

Like the tale of Jonah living happily inside a whale without suffering any physical discomfort, there's something fishy about the universe popping out of nothingness. And not just because in everyday experience we do not observe kittens or lawn furniture magically materializing. The problem lies deeper. It's simply that even if this narrative is true, the "magically materializing" business is really no explanation at all.

So let's back up to be strictly honest about what we know and don't know. We can begin with truths no one can dispute, the way René Descartes did when he said, "I think, therefore I am." Our absolute bedrock bottom-line reality is not that we humans descended from plankton on a world born near a third-generation star 4.65 billion years ago. That may seem certain to many in our modern world, but here's an even more inarguable starting point: *We find ourselves to be conscious, in a matrix we call the universe.*

We seek some understanding or larger context for this existence. If we find theological models inadequate, we turn to science, whose researchers state, once again, that the universe popped out of nothingness by some unknown process. They go on to "explain" that life eventually arose equally inexplicably. And this life manifests individual awareness that itself is enigmatic.

This is the scientific explanation for what's going on.

No wonder, in many circles, that such elucidation is not regarded as superior to the old-fashioned "God did it."

This is not to blame science in any way. Far less than one-trillionth of 1 percent of the cosmos lies within view of our telescopes. And even *this* is just a small fraction of the actual cosmos, because the majority of everything is composed of unknown entities. Our sample size is thus minuscule. Moreover, increasing evidence indicates that the universe may be spatially infinite (more on this in chapter 18). That would make it infinite in its inventory as well, in which case everything that lies within view is actually *zero percent of the whole universe*, as any fraction of infinity is nothing. The point is, if we're to be honest, our data are currently too negligible to allow valid generalizations. The sample size is simply too small to be trustworthy.

Sadly, this fact is rarely, if ever, acknowledged, especially on TV science programs. Discussing our lack of information would constitute "dead air" that would motivate no commercial sponsor.

Yet, in truth, we recently discovered the universe is mostly composed of dark matter, but we don't know what it is. Then we discovered that, actually, it's mostly dark energy, but we don't know what *that* is, either. Dark energy's existence was postulated because in 1998, we found the universe's expansion, which was always believed to be slowing down, was actually mysteriously speeding up. Dark energy is apparently some kind of antigravity force that's blowing the cosmos apart.

We also have no idea how self-replicating life began. Moreover, we find ourselves in a universe fine-tuned for life but have no idea how—except by speculating an infinity of universes in which we are the lucky ones.

Given this vast absence of hard data, cosmologists try to compensate with a reliance on *models*, with guesses about starting conditions and intermediate events. This still would not be a problem if people didn't take them so seriously—if they realized that these are just *starter* models.

In the early twenty-first century, these models include catchy notions intended to impart a picture of the cosmos, even if they lack supporting evidence. In scientific language, concepts like cosmic membranes and string theory are *nonfalsifiable*—they cannot be proven or disproven. They will almost certainly be abandoned or greatly modified in our lifetimes, replaced by other models that will eventually be discarded, too, just as the "slowing-down universe expansion" of 1997 was replaced by the "speeding up" model of 1998.

Thus, to address that seven-year-old truthfully would be to confess that *science cannot currently answer the simplest questions about existence.*

True, cosmologists speak of the "2.73-degree Kelvin cosmic microwave background" and the "13.8 billion years since the Big Bang," and these seemingly precise figures, complete with decimal points, create a verisimilitude of credibility. The models are then stated repeatedly, and this repetition itself endows them with a substantive aura. But this doesn't mean that they are, in fact, hard truths.

Happily, the preceding, gloomy-seeming overview of our present state is not the end of the story. It is actually only the beginning. Because there exists an alternative model for What All This Is.

The alternative is necessary because modern cosmology, in its attempts to explain the cosmos, keeps committing an odd oversight: It scrupulously holds the living observer at a distance from the rest of the universe. It asks us to accept a dichotomy, a split.

In this corner stands us, the living—the perceivers of it all. And in the other corner lurks the entire dumb universe, slamming into itself via random processes.

But what if we are linked? What if the whole insensible model can suddenly make sense by putting everything together? What if the universe—nature—and the perceiver are not stand-alone entities? What if one plus one equals . . . one! And indeed, what if the past century of scientific discoveries point compellingly in this very direction—if only we are sufficiently open-minded to see what it tells us?

In reality, the clues never stop arriving. In February 2015, the *New York Times* ran a story on "Quantum Weirdness," which it subtitled, "New Experiments Confirm That Nature Is Neither Here Nor There." Yet neither the clearly puzzled author, nor many readers in all likelihood, smiled to themselves and thought, *Of course! That's because nature is indeed both here AND there.* When you try to locate it solely in either place, you end up with paradoxes and illogic.

Quantum theory found a connection between consciousness and the nature of particles nearly a century ago. Yet we've ignored this, or come up with dizzying explanations involving an infinite number of alternate realities.

Discovering what's real is actually a happy endeavor. It requires us to walk the labyrinthine hallways of the most intriguing twenty-first-century science concepts and to examine existing ones afresh. Exploring astounding things like time and space and how the brain works would promise to be an enjoyable excursion even if it were a mere aimless diversion, a Sunday stroll.

But as we shall see, both the voyage to a clearer picture of the cosmos, and the ultimate destination itself, are more than eye opening. They are fun.

THE SEVEN-MILLENNIUM QUESTION 2

> The day which we fear as our last is but the
> birthday of eternity.
> —Lucius Annaeus Seneca, "De Brevitate Vitae" (c. 48 C.E.)

In attempting to tackle the fundamentals about ourselves and the universe, we usually turn to the science of cosmology, although some continue to embrace religious explanations. But those who find neither avenue leading to their desired destination can consider a very different model of reality. This fresh paradigm, far from abandoning science, uses discoveries published since 1997, and reexamines others that unfolded even earlier.

Before we plunge into this new adventure, however, it's helpful to see what the great thinkers have already come up with through the ages. We don't want to reinvent the wheel if it's already there.

This requires that we overcome our biases of ethnocentrism and modernism. That is, we often reflexively assume that our

Western culture, and people alive today, have a superior grasp on deep issues compared with foreign civilizations and those who lived before us. We base this on our advanced technology. Those poor slobs a century ago had no indoor plumbing, window screens, or air conditioning. Could anyone have deep insights when sweating in a sticky bed and beset by droning mosquitoes? Could they conjure profundities while tossing their night wastes out the window each morning?

Thus it may surprise anthropology students to learn that vast areas of human knowledge commonly grasped by the educated classes of the eighteenth and nineteenth centuries are greeted today with blank stares. It's therefore not true that twenty-first-century teenagers have more knowledge than their nineteenth-century analogs—just *different* knowledge.

Every farm boy in 1830 knew precisely how the sunrise shifts its weekly rising and setting points and could identify the songs of birds and the detailed habits of the local fauna. By contrast, very few of our friends or family members today are even dimly aware that the Sun moves *to the right* as it crosses the sky daily. Confessing such ignorance about something so "sky is blue" basic would have been met by disbelief in the nineteenth century.

To be sure, some areas of knowledge have thoroughly eluded all humans, present and past alike. For example, we've proven ourselves chronically deficient at foreseeing the future—even anticipating conditions a few decades ahead. No genius of the classical Greek period, no great writer in global literature, no passage within any religious text ever suggested that there exist tiny creatures too small for the eye to perceive, let alone that such *germs* are responsible for most of the diseases that plague us. Before 1781, no one suspected that perhaps there might be additional planets beyond the five bright luminaries known since the Neanderthals. Until just a few centuries ago, no one suggested that blood circulates through the body, or that the air we breathe consists of a mixture of gases rather than a single substance. Thus, for all the New Age or religious malarkey extolling the supposed accuracy of ancient "prophecies," the actual track record is worse than dismal.

We have done no better in modern times. The futurists who helped prepare the 1964 New York World's Fair depicted typical homes of the year 2000 as having flying cars and personal robots. In popular literature and cinema, the 1968 classic *2001: A Space Odyssey* showed lunar colonies in the year 2000, and a Jupiter voyage with a human crew a few years later. The 1982 cult favorite *Blade Runner* depicted Los Angeles in 2019 as being relentlessly rainy from an implied climate change that turned California into a chronically wet place. That city was also crammed with ultra-tall buildings and flying police cars. No futurist during the hippie years foresaw today's ubiquitous cell phones, body piercing, or the super-fast modernization of China.

The point is, our present level of perspicacity seems no better than it was a few centuries ago. Nor is it worse. And when it came to pondering our place in the universe, our ancestors were at least as obsessed as we are. So, given that the vast majority of humans who ever lived are *not* alive today, it would be an oversight to ignore their insights.

Rather than assuming our ancestors were too backward to think deep thoughts, or going the other way and idolizing past civilizations as being supernaturally in sync with nature, let's look at the actual written record.

It is not necessary to summarize the bedrock beliefs of every civilization. Certainly in the Western Hemisphere, if we're to begin our account seven thousand years ago, even before the invention of the wheel, the worldview was consistently dominated by a time-based obsession with the afterlife. This in turn revolved around appeasing the gods—like the Egyptian sun god Ra, creator god Amun, and mother goddess Isis.

Here, the earliest writings showed no interest in solving nature's mysteries through observation or logic. Instead, magic and superstition ruled. One of the authors found a primitive hieroglyphic example from forty-seven centuries ago, inscribed on the subterranean walls of the lonely pyramid of King Unas at Saqqara in Egypt. This 2006 visit had been guarded against terrorists by a jeepload of heavily armed troops—all to observe glyphs that were not exactly Deep Thoughts.

They were magic spells featuring a "mother snake."

From there, in the twenty-seventh century B.C.E., literature had nowhere to go but up. But it took a thousand years before incantations, grain tallies, and long-winded accounts of the everyday goings-on of the Pharaoh's family gave way to genuine insight. The oldest religious text, the Sanskrit *Rig Veda* from around 1700 B.C.E., pondered "the Sun god's shining power" and said, poetically, "Night and morning clash not, nor yet do linger." Translation: Stuff happens.

By the time the Old Testament books were penned a millennium later, a key point was a stationary Earth ruled by a single, easily upset God. The rabbis of the time showed no inclination to question this prevailing worldview. They duly filled the pages of Genesis and Deuteronomy with the flat-earth, glued-in-place mindset of their time, with a strict dividing line between us mortals below and heaven above. Figuring out how nature operated was on nobody's to-do list. Indeed, the things that provoke our curiosity today—the nature of life, and time, and consciousness, and the working of the brain—all would have seemed alien to early civilizations. Everyday survival was priority number one, behaving according to Scripture so that God wouldn't smite you was number two, and debating issues like whether space is real never made it to the campfire agenda.

Back then, everyday life's main illumination was the Sun and Moon, and just to make sure everyone was paying attention, these lights kept shifting position. They repeated their dog-and-pony show daily. Despite lacking any inclination to explain the natural world around them, the ancient scribes couldn't ignore light—so central to every aspect of life—so they emphasized this topic in the opening lines of Genesis. Of the first one hundred words in the Bible, fully eight are either "darkness" or "light."

(They may have been onto something. We will see, in our own explorations, that light, or at least energy, is indeed a central character in Reality's puzzle.)

In that era, no one had a handle on the actual structure of the cosmos, how we perceive it, or how everything might be linked. There was insufficient information. Then, as now, people didn't want to spin their wheels on topics that went nowhere.

But repetitions were another story. They stirred the intellect. Our brains are built to notice patterns. We readily link them with others. If the phone rings just as we sit down to dinner for six nights in a row, this isn't going to escape our attention.

The most prominent pattern involved that blinding ball of fire. It always crossed the heavens from left to right. It faithfully rose in the east. On the incomprehensible side the Sun was obviously a god of some sort. Probing its secrets surely seemed mission impossible.

Yet "figuring stuff out" became a priority on the sunny islands of Greece some six centuries before the birth of Christ. More to our point, it opened the doors to the earliest realistic contemplations about our place in the universe. It happened because, for the first time, rationality competed with magic. Observation and logic were prized at long last.

Logic involves cause-and-effect sequences. A causes B, which then causes C. Everyone comes running from the fields after a goat shed collapses because an olive tree fell on it. The tree was knocked over by the wind. This happened at midday when the wind usually blows strongest. One of the village's smarter men connected A with C and wondered aloud: Might the hot overhead Sun be the wind's instigator? Hey, this was fun—uncovering a possible link between the Sun and a dead goat. The Greeks fell in love with this newfound tool of logic.

They were on the right track, but the very early Greeks— the first true practitioners of science—reached stumbling points fairly quickly. Two thousand years later, in the early seventeenth century, Italian physicist Evangelista Torricelli did indeed explain why the wind blows, and it *did* involve the Sun. But the ancient Greeks were hampered by their need to keep their gods in the picture. So, why did the god of the west wind, Zephyrus, choose to blow at some times but not others? The villagers would shrug; the gods had their own inscrutable reasons.

If the goat was dead, Zephyrus was apparently punishing the goat herder for some transgression. Guessing the crime even became a favorite neighborhood gossip topic. Infidelity was always a good bet, although hubris could often be suspected. You couldn't understand divine motives, so why bother trying to figure out anything? In particular, a "first cause"—what starts the ball rolling—was vexingly impossible to pin down.

Yet even if cause-and-effect rationality reached blank walls quickly, the early Greeks admirably didn't quit. And like science even today, especially the quantum theory experiments we will explore later, the ancients had to deal with *verisimilitude*, a wonderful word that means "the appearance of truth."

Something that appears true may indeed be true. Or it may not be. The Sun crossing the sky while Earth remains motionless is a verisimilitude, an appearance. It *seems* true. It still appears true today, which is why we say "the Sun is setting" and not "the horizon is rising." It was an amazing leap for Aristarchus on the island of Samos, fully eighteen hundred years before Galileo, to insist that you'd observe the same effect if it was Earth that was spinning while the Sun was stationary—and that this made more sense because the smaller body should logically revolve around the larger one.[1]

---

[1] Being far ahead of everyone else in the world, especially concerning some fundamental facet of life, has rarely bestowed any benefit. Who knows the name Aristarchus today? We checked; there isn't a single high school named for him in the United States. At least he wasn't put to death, unlike many other pioneers in thinking.

We will try to remember this idea of verisimilitude later, when we, too, are faced with alternative ways of interpreting everyday observations.

Meanwhile, Aristotle, in his groundbreaking *Physics*, held the view that the universe is a single entity with a fundamental connectedness between all things, and that the cosmos is eternal. You needn't get hung up in the cause-and-effect business, he argued in the fourth century B.C.E., because everything has always been animated and has a kind of innate life or energy to it. There is no starting point. Actually, Aristotle hardly went out on a limb to say these things, as this solipsistic view had many adherents before he arrived on the scene.

Aristotle didn't quit there. In Book IV of *Physics*, he argued that time has no independent existence on its own. It only subsists when people are around; we bring it into existence through our observations. This is very much in line with modern quantum experiments. No physicist today thinks that time has an independent reality as any sort of "absolute" or universal constant.

Still, neither Aristotle, nor Plato, nor Aristarchus, could abandon the dichotomy of us mortals existing here below while above us dwelled a parallel heavenly realm inhabited by the gods.

But things were very different in the East. Even before the Roman Empire, which retained the Greek gods (albeit with new names), a main branch of South Asian thought was being codified in texts such as the *Bhagavad Gita* and the Vedas. Their model of reality, soon known as Advaita Vedānta, was astonishingly unlike the Western worldview.

In common with Aristotle, Advaita taught that the universe is a single entity, which it called Brahmin. But unlike the Greeks, this "One" included the divine, as well as each person's individual sense of self. All appearances of dichotomy or separateness, it insisted, are mere illusions, like a rope being mistaken for a snake. Advaita Vedānta went on to characterize this One as birthless and deathless, and essentially experienced as consciousness, a sense of being, and bliss.

Moreover, the Advaita teachers averred, realization of this was *the* goal of life. Not appeasement of gods, nor contributions to clergy, nor even any concern for an afterlife, but merely awakening to a full grasp of reality. Later spin-off religions such as Buddhism and Jainism retained these fundamentals. Today, the world still remains essentially divided into these basic two views of reality, Western and Eastern, dualistic and non-dualistic, that existed over a millennium ago.

The Eastern religions maintain that some individuals through the centuries have periodically enjoyed the "enlightenment" experience. That is, they awoke and saw the truth, and were swept into ecstasy and a sense of freedom.

A fascination with such Eastern views arrived in Western countries in the late nineteenth century, abetted by visits of a succession of influential, articulate Indian teachers such as Paramahansa Yogananda, Swami Vivekananda, and more recently Deepak Chopra. In the 1940s, Yogananda, through books such as his best-seller *Autobiography of a Yogi*, attempted to justify the Eastern view of the cosmos through science. By most accounts, such efforts sounded forced and the science arguments were less than compelling. They probably persuaded only those who were already on board.

But the quest itself was noble. If a person seeks knowledge of reality and one's nature and one's place in the universe, what if she has no spiritual calling? What if she solely demands fact-based evidence? Can these deep issues be tackled decisively by science alone?

That is our sixty-four-thousand-dollar question—and the real starting point for our journey.

IN THE BEGINNING . . . 3

All is change; all yields its place and goes.
—Euripides (c. 416 B.C.E.)

No matter what picture of the universe one embraces, time seems to play a key role. Indeed, our existing models are so thoroughly time based, they can neither be understood nor disproved without also understanding time itself. Thus we must tackle it before anything else.

This is no mere philosophical matter. It goes to the heart of our perceptions and lies at the fulcrum between the observer and nature. Certainly, we use time constantly. We make appointments and look forward to vacation plans, and some of us fret about the afterlife. If there is one big difference between people and animals, it is not that we are unafraid of vacuum cleaners. It is that we are time obsessed.

On one level, what we commonly mean by time is inarguably real. Our car's GPS announces that if we stay on this highway we will reach Cleveland in 3 hours and 48 minutes. And we do. Moreover, while we do that, countless other events unfold in our bodies and elsewhere on the Earth.

Yet this agreed-upon interval is, on closer inspection, as fishy and intangible as the question of what exactly happened at midnight on New Year's Eve.

The question of time has tormented philosophers for millennia, and this torture shows no signs of abating. Happily, unlike the intricacies of, say, Middle East politics, here we have only two contrasting viewpoints.

One is the opinion held by such noted smart people as Isaac Newton, who saw time as part of the fundamental structure of the universe. He believed it to be inherently real. If so, time constitutes its own dimension and stands separate from events, which unfold sequentially within its matrix. This is probably how most people view time.

The opposing view, argued for centuries by other smart people such as Immanuel Kant, is that time is *not* an actual entity. It is not a kind of "container" that events "move through." In this view, there is no flow to time. Rather, it's a framework devised by human observers as they attempt to give organization and structure to the vast labyrinth of information whirling in their minds.

If this latter view is true, and time is only a kind of intellectual framework along the lines of our numbering systems or the way we order things spatially, then it certainly cannot be "traveled," nor can it be measured on its own.

This means that clocks do not determine or keep track of time, but merely offer evenly spaced events as one digital number is replaced by another, or a minute hand is now here and now there. While these events proceed, other reliable rhythms simultaneously unfold elsewhere. And, of course, the lengths between each tick and tock are arbitrary, having been agreed upon by human council rather than some decree of nature.

The tick-tock idea began with Sun-based changes observed by people occupying a far more outdoorsy world than today's. Sumerians and Babylonians more than six thousand years ago utilized the concepts of "day" and "year" and "month." Soon after, the ancient Hindus defined specific units of time such as the *kālá*, which corresponds to 144 seconds.

The Hindus created a dizzying variety of intervals. At either end of their time spectrum the units were so extreme, they were useless in practical terms—and close to incomprehensible. These included the *Paramaṇu*, with a length of about 17 millionths of a second, and the *Maha-Manvantara*, which is 311.04 trillion years. Their long-interval units meshed with their creation and destruction myths, in which the cosmos undergoes cycles of clarity alternating with periods of human darkness, each called a *yuga*.

More practically, the ancient agrarian world relied on seasonal ways of reckoning, and these cycles were determined with amazing accuracy in civilizations like the Maya. Smaller units than months and days trickled into everyday usefulness, first with the creation of the dripping-water or falling-sand hourglass, and later the discovery of the pendulum effect by Galileo Galilei. In 1582 he noticed that the chandeliers hanging from long chains in the Piazza del Duomo kept swaying back and forth in the same period regardless of the swing's amplitude, and—following an impressive bit of procrastination—wrote about this in 1602. This effect, experienced by children in playgrounds, amounts to the fact that when a parent gives a child a strong push, the swing's period of travel from one end of its oscillation to the other is no different from when she is just sitting quietly with the swing barely moving at all.

The period is basically determined by the length of the chain, a property called *isochronism*. It turned out, a string or chain 39 inches long produces a back-and-forth period of exactly 2 seconds. It wasn't long before this principle was utilized in grandfather clocks, whose long metal rods, just over 6 feet, ticked off near-perfect seconds.

Portable timekeeping took a leap with the invention of the balance spring watch in the second half of the seventeenth century, thanks to breakthroughs by Robert Hooke and Christiaan Huygens. Then accuracy skyrocketed after the 1880 discovery by the Curie brothers, Jacques and Pierre, that quartz crystals naturally vibrate when a bit of electricity is applied to them. If cut to a particular size and shape, they'll reliably oscillate 32,768 times a second, which is a "power of 2"—it's 2 multiplied by itself 15 times over. An electronic circuit has no trouble counting these oscillations and thus marking off evenly spaced seconds. This ultimately made precise portable timepieces—the quartz movement still utilized today—cheaply available beginning in 1969. With everyone now able to agree on the "right time," the busy modern world with its appointments and scheduling settled into a shared, time-focused reality.

Through it all, however, the fact of pendulum swings, mechanical balance beam oscillations, and quartz vibrations was still no evidence of time. They all merely provided regular repetitive motions. One could then compare some repetitive events with others. One could notice, for example, that while a grandfather clock pendulum makes 1,800 swings, a candle might burn down 1 inch, and Earth would turn one-forty-eighth of a full rotation. Certainly, one could call the elapsing of all these events "a half hour," but that didn't mean that the time period had some independent reality, like a watermelon.

Then the whole business suddenly grew much odder with the discovery that some events could start unfolding faster than they had before, relative to others. Things started to become seriously disconcerting with Einstein's strange but grudgingly logical ideas that he incorporated into both his special and general relativity theories of 1905 and 1915, respectively. In them, Einstein elaborated on and explained curiosities and paradoxes noted in the preceding decades by George FitzGerald and Hendrik Lorentz. In a nutshell, a totally unexpected revelation emerged: Even if time is an actual entity, it cannot be a constant like lightspeed or gravity. It flows at different rates. The presence of a gravitational field retards the passage of time, as does rapid motion.

We're intuitively ignorant of this because we all attended a high school where everybody hung out in the same gravitational field—and never, even in our wildest teenage years, sped our car in a joyride faster than an eight-millionth of the speed of light. Because one must go 87 percent of lightspeed to feel time slow by half its normal rate, we've never even come close to directly experiencing time's fickleness—a function of our still-sluggish ground vehicles rather than any personal wisdom.

Astronauts do better. Orbiting at one-twenty-six-thousandth the speed of light, they can actually gauge the amount by which their time runs slow, using sensitive clocks—which brings up a seldom discussed puzzle. Though they move faster, astronauts have also traveled away from Earth's surface into a weaker gravity, which has the opposite effect, speeding their passage of time. Turns out, their high-speed factor prevails. They age *less* quickly than people on the ground. They'd have to be eight times higher than the International Space Station's orbit, or two thousand miles above Earth's surface, before the weaker gravity there exactly balanced their now slower orbital speed to let them age at the same rate as those back home. Still farther away, timepieces on the Moon tick faster than those at mission control in Houston—even if nobody compensated Apollo crews with early Social Security benefits.

These time distortions aren't subtle, nor are they merely of academic interest. Those GPS satellites simply wouldn't work if continual compensations weren't added for various time-warping effects. Since receiving precise time signals from each satellite lies at the very heart of that navigation system, anything that throws off the instruments' or receivers' time passage will blow the whole thing.

Are you a truly nerdy, geeky person who cares about such technological or physics details? If so, consider the many wrinkles in how time seems to flow, all introduced by the very technology designed to measure it:

Wrinkle one: Satellites travel at 8,700 miles per hour, slowing their clocks.

Wrinkle two: They're distant from Earth in a reduced gravitational field, which accelerates their time relative to Earth's surface.

Wrinkle three: GPS users on the Earth's surface are located at various distances from Earth's center (at Denver's high altitude versus low-altitude Miami, say), producing a variety of time-passage rates.

Wrinkle four: The difference in Earth's rotation speed at separate ground-based locations produces inconsistencies in their agreement about the passage of time, which is called the *Sagnac effect.*

Wrinkle five: Time runs slower for all earthly observers (as compared to any future lunar colonists) because of our planet's 1,040-mph equatorial spin. (The speed decreases the farther one is from the equator.)

Wrinkle six: Satellites' time passage continually changes because their slightly elliptical orbits make them speed up and slow down, plus they zoom through irregularities in Earth's gravitational field due to things like our planet's equatorial bulge.

All told, six separate Einsteinian time distortions affect receivers' clocks; half of these also distort the satellites' clocks. They must all be accurately and continuously corrected. Any inconsistencies would ruin the system's accuracy, big time.

And always remember: We're not talking about the warping of an actual entity called time. We're noticing only that events unfold at more leisurely rates, or more hurriedly, than they did before, *relative to others.* This remains a central point. A hawk flaps its wings slowly, whereas a hummingbird's wings beat furiously. Sure, we could bring our concepts of time into the discussion, yet we needn't do so. The event is one thing. How we categorize or measure it is another.

For those who may imagine that such "time warps" are only a mind game, a mere theory, the fact is, Einstein's time dilation even causes death. When cosmic rays (highly energetic particles striking our atmosphere) collide with molecules in the upper layer of air, they break atoms apart like a cue ball smashing a stack of billiards. The resulting rain of subatomic particles includes some that can be lethal to humans if they strike the wrong bit of genetic material. These *muons* dash through our bodies constantly, causing some of the spontaneous natural cancers that have always plagued our species. Over 200 of these penetrate each of our bodies every second—more if you live higher up, like in dangerous Denver again. The point is, muons, intermediate in mass between protons and electrons, exist for just 2 microseconds before decaying into harmless by-products. And a few microseconds is not long enough for them to make it all the way to Earth's surface and into our cells, even though they travel a hefty fraction of the speed of light.

Muons should decay so quickly after being created thirty-five miles up, they ought not be able to reach us. They should never arrive here. They should not cause us any trouble. But they do. What we count as a few microseconds becomes a longer period of time to the muons. Long enough to live on and on. Their time has slowed because of their high speed. To us observing it, the muon's life has been extended—and ours perhaps shortened. Yet from the particle's perspective, time passes normally.

There are places in the cosmos where a million years of events pass while a single second's worth of activities simultaneously elapses here on Earth. Yet both feel a normal passage of time.

So observers in different places experience out-of-sync sequences. If the rate of the passage of events depends on factors like the local gravity and one's speed, how can there be a stable commodity called time?

Exploring this, physicists look to see if time is critical, or even has existence, in their physics equations—or whether what has been spoken of as time is merely the fact of *change*, long represented by the capital Greek letter delta: Δ. Doing so, they find that Newton's laws, Einstein's equations in all his theories, and even those of the quantum theory that came later, are all time symmetrical. Time simply plays no role. There is no forward movement of time. Many in the physical sciences thus declared time to be nonexistent.

As Craig Callender wrote in 2010, in *Scientific American*[2]:

> The present moment feels special. It is real. However much you may remember the past or anticipate the future, you live in the present. Of course, the moment during which you read that sentence is no longer happening. This one is. In other words, it feels as though time flows, in the sense that the present is constantly updating itself. We have a deep intuition that the future is open until it becomes present and that the past is fixed. As time flows, this structure of fixed past, immediate present and open future gets carried forward in time. This structure is built into our language, thought and behavior. How we live our lives hangs on it.
>
> Yet as natural as this way of thinking is, you will not find it reflected in science. The equations of physics do not tell us which events are occurring right now—they are like a map without the "you are here" symbol. The present moment does not exist in them, and therefore neither does the flow of time. Additionally, Albert Einstein's theories of relativity suggest not only that there is no single special present but also that all moments are equally real.

Philosophers generally agreed. After all, the past is just a selective memory; *your* recollections of an event are different from mine. Both memories are simply that—signals from brain cells, neurons firing in the present moment. If the past is an idea

---

[2] http://www.scientificamerican.com/article/is-time-an-illusion/.

that can only occur in the here and now, and the future is also just a concept happening strictly in the present, there seems nothing but now. Always. So is there *really* a past and a future? Or just a continuum of present moments?

This debate is not new. As we've seen, several classical Greek writers believed that the universe is eternal, with no origins at all. Possessing such an infinite past with no beginning made time seem meaningless. Eternity, after all, is fundamentally different from "time without end." Even as long ago as the fifth century B.C.E., Antiphon the Sophist, in his work *On Truth*, wrote, "Time is not a reality, but a concept or a measure."

In the town of Elea, Parmenides seconded this in his poem, *On Nature*, in a section titled "The Way of Truth," in which he stated that reality, which he referred to as "what-is," is one, and that existence is timeless. He called time an illusion.

Soon after, still in the fifth century B.C.E, in that same Greek town of Elea, the famous Zeno created his enduring paradoxes, which in the next chapter will provide critical instruction on how to tell the difference between the conceptual realm of ideas and math versus the actual physical world. (This will resolve that old nagging paradox of the tortoise racing the hare, which has been filed in your brain all these years in a section devoted to "miscellaneous mental torment.") Zeno will also help show us how neither time nor space are actual physical entities.

In sharp contrast to the carefree Greek musings on eternity, medieval theologians and philosophers tended to see God alone as infinite. To them, His creation, the universe, must therefore indeed have a finite past, a specific moment of birth, and an assumed expiration date. By this reasoning, time *is* part of the cosmos and thus is itself finite.

Enough philosophizing. Though such debates continue today, they've been offered only to illustrate how time's reality, so assumed by the public, continues to be seriously doubted among people with excessive leisure time who ponder such things. More central for us, it is doubted even in the mainstream of science. And it is the science alone that we will now continue to pursue as we heat up our hunt for a definitive resolution to the time business—our first key to understanding existence, death, and our true relationship with the cosmos.

We must shift to the only place in science where a directionality of time is assumed to be needed: the field of thermodynamics, whose second law involves a process called *entropy*. This natural inclination to go from order to disorder necessitates an "arrow" or direction to time. If such an arrow exists, then time is a real item after all and will disconcertingly tick away the remaining minutes of your life.

We'd better hurry up and get to the bottom of this. We'll call on real people who helped clarify what's going on. This odyssey will lead from Parmenides and Zeno, whose world was very different from ours, to nineteenth-century Europe and a name known to every physics student—the brilliant, fascinating, but ultimately tragic Ludwig Boltzmann.

ZENO AND BOLTZMANN

4

Life . . . presupposes its own change and movement,
and one tries to arrest them at one's eternal peril.
—Laurens van der Post, *Venture to the Interior* (1951)

We should probably begin with Parmenides, who was born around 515 B.C.E. in Elea, on the Greek mainland. He is known for founding Eleaticism, which quickly became one of the leading pre-Socratic schools of Greek thought. But though only small fragments of his principal work—the lengthy, three-part *On Nature*—survive, there's really no need to complicate what is essentially a simple worldview, one that very much jibes with biocentrism 2,500 years later.

Parmenides' views were seconded and championed by Zeno, born in the same settlement twenty-five years later. Both men tirelessly argued that the apparent multiplicity of objects we see around us, along with their changing forms and motions, are but an appearance of a single eternal reality they called "Being." This was actually very much in sync with what had been written in Sanskrit texts a thousand years earlier, although Parmenides and Zeno seem to have arrived at their perceptions independently.

The Parmenidean principle boils down to "all is one." This may seem like idle philosophy, but it's pregnant with vast experiential perceptions that affect everyday experiences then and now. A babbling brook, for example, would be apprehended by the Eleatics as an expression of the limitless energy and play exhibited by *Being* or existence, whereas the opposing school (almost universally embraced in our modern time) is that a multiplicity of separate, quasi-independent objects like water molecules and pebbles are exhibiting cause-and-effect-derived actions in a space- and time-based matrix in which these disparate items come and go individually. And although the multiple-causation versus the "single animated essence" views may at first seem philosophical and unimportant distinctions, each in turn leads to very different conclusions about what's actually unfolding and what kind of reality we're part of. It's actually a life-changing topic.

Perhaps that's why Parmenides and Zeno, almost obsessively embracing their stone-simple concept of Being, felt a kind of Paul Revere–like need to spread the word. Doing so, they insisted that their view didn't require faith or perception but could be proven through logic. Because they said that all claims of change or of non-Being are illogical, Zeno in particular created a series of paradoxes designed to disprove all time- or motion-based arguments, which he maintained would lead inexorably back to the simplicity of the One Energy. Even today, Zeno's paradoxes are taught, debated, and still generally considered valid.

More than that, Aristotle admiringly credited Zeno as being the inventor of the *dialectic*, a word that later became synonymous with formal logic. This was ironic in a way, since Zeno's entire purpose was to support and recommend the Parmenidean doctrine of the existence of "the one" indivisible reality, which is about as unconvoluted a position as is humanly possible to take. So in looking at Zeno's paradoxes, we should always remember that their goal was not to be clever or to pull the rug out from under the machinations of logical thought, but to contradict and disprove the widespread belief in the existence of "the many"—meaning individual objects with distinguishable time-based qualities and separate motions.

Zeno created many paradoxes to prove his point, but we'll only list the three best known. Probably everyone has heard of the Achilles and the Tortoise tale, called by various other names as well. It starts by letting the slower-moving tortoise have a head start, and then Achilles attempts to catch up and pass it in a race. Let's say the tortoise goes half the speed of Achilles. Well, as soon as Achilles reaches the place where the tortoise was positioned at the outset, the tortoise has meanwhile moved on by half that distance. When Achilles reaches this new position, the turtle has meanwhile slowly advanced to yet another new position, halfway beyond its initial advancement and Achilles' new position. And when Achilles attains the new tortoise position, there's no avoiding the fact that the animal has managed to move ahead by another half of *that* distance. The halves keep getting further halved, but Achilles can never catch the tortoise.

A second paradox is similar: If Homer wants to reach a man selling grapes from a cart, he must first advance to half the distance between his front door and the fruit vendor. Then, he must arrive at a point that is half of that distance. Then half of *that*. It's obvious that half of the remaining distance will always have to be attained first, and this creates an infinite task that has no conclusion. Homer can never buy the grapes.

Our third paradox involves an arrow in flight. Obviously, at any given instant in time the arrow must be *somewhere* and nowhere else. It is no longer where it used to be, and it is not yet at its next possible point in its flight. In other words, at every instant there is no motion because the arrow is exclusively at one precise position and thus at rest. If everything is motionless at every instant, and time is entirely composed of instants, then motion is impossible.

At each moment we are at the edge of a paradox known as "The Arrow," first described 2,500 years ago by Zeno of Elea.

Since nothing can be in two places at once, he reasoned that an arrow is only in one location during any given instant of its flight.

But if it is in only one place, it must momentarily be at rest. The arrow must then be present somewhere, at some specific location, at every moment of its trajectory.

Logically, then, motion per se is not what is really occurring.

Rather, it is a series of separate events.

The forward motion of time—which the movement of the arrow is an embodiment—is not a feature of the external world but a projection of something within us, as we tie together things we are observing.

By this reasoning, time is not an absolute reality but a feature of our minds.

In our busy lives, there may be a tendency to dismiss such logic as mere puzzles, and brush them off as if shooing away a fly. But the greatest minds through the centuries have been tormented by Zeno's paradoxes. Although some have grandly announced "solutions," the consensus is that they're still valid today. The paradoxes can actually be solved by biocentrism. By seeing that time and space are not actual commodities like coconuts, biocentrism says they cannot be divided in half again and again to produce such conundrums. Alternatively, one might see that the physical world is not the same as the abstract mathematics or even simple logic we might use to describe it. Logic demands symbolic thinking, where objects and concepts are represented by ideas, whereas the actual world doesn't have to play by those semantic rules. By this reasoning, Zeno's paradoxes arise because we've switched between the physical and the abstract. Since we're so rooted in our thinking minds, we've forgotten how to recognize the difference. In the abstract world, those endless halvings are a stopper and prevent Homer from ever buying the grapes. But in the actual nonsymbolic reality of nature, he can simply walk over and hand the vendor a drachma.

For our purposes, however, it's enough to show that space and time—the seemingly bedrock grid many of us assume to be a real framework for the universe—are fragile mental constructs whose logical existence can be shaken by the likes of Zeno. If he's right and motion cannot actually exist, what is it that we experience when we watch a home run ball narrowly miss the foul pole? What's going on there? Before we get to that, we have one more task in our demotion of time: to see if any area of science can support it.

This takes us to Austrian physicist and philosopher Ludwig Boltzmann, who was born in 1844. Beginning his study of physics when he was nineteen at the University of Vienna after his father died, he earned his PhD at age twenty-two and became a lecturer. It was a heady time for physics, and Boltzmann was particularly fascinated with developing a way to statistically figure out how to explain and predict the motion and nature of atoms, which let him accurately determine such properties of matter as viscosity—basically how gooey or runny liquids are.

Boltzmann struggled his whole life with wild swings in mood, which flowed like his beloved fluids at vastly different rates. Today he'd probably be diagnosed as suffering from bipolar disorder. It often made his relationships with his colleagues difficult, but it didn't prevent him from making major advancements in explaining how matter behaves. In doing so, he was in a way anticipating the quantum mechanics that would arise decades later, which also rely on statistics to understand how the physical world operates. Before finally succumbing to depression and hanging himself at the age of sixty-two, he created three laws of thermodynamics, of which the second—commonly associated with the idea of entropy—remains the most famous.

Entropy enters our own reasoning because it is the single area of physics that seems to argue for the existence of time. In all others, whether the equations of general relativity, or Kepler's laws of planetary motion, or quantum mechanics, everything is time symmetrical—stuff happens, but there is no external arrow or directionality that makes time an actual entity.

Boltzmann created a model of atoms in a gas that resemble colliding pool balls. He showed that if they're all confined in a box, each collision causes a distribution of velocity and direction that becomes increasingly disordered. Ultimately, even if a high degree of order was the initial condition—say one side of the box contained hot, fast-moving atoms and the other side cold, slower-moving ones—this structure would vanish. Such an ultimate state of large-scale uniformity, or total lack of order even on the microscopic level, is called *entropy*. Given enough time, the final condition—a state of maximum entropy—is thus inevitable.

Notice the word "time" was central to the process. And that's the point. The act of going from structure to disorder, of increasing entropy, is a one-way process. The eventual uniformity, and the obliteration of all temperature differences, appears to be time based because it's not reversible. We see this in everyday life. The drawer in which we keep our socks never somehow gets more arranged, with matching pairs increasing their frequency no matter how long we rummage through them. Disorder happens naturally. And if this really is physical or mathematical evidence for a "direction" or "arrow" of time, then time is real.

Arrows of time are not taken lightly in physics. Stephen Hawking once argued that if the universe ever stops expanding and begins to collapse, the arrow of time would point in the opposite direction and physical processes would reverse themselves on every level. Presumably we'd never notice anything amiss, since our own mental workings and brain functions would be running backward, too. In any case, Hawking eventually decided that reversed time couldn't happen, and he changed his mind as if to demonstrate the process.

We have no other hard evidence for time except for Boltzmann's second law of thermodynamics. But this entropy is no small thing. It's pretty inarguable. Is there any way out that can make us not seem like naive pleaders as we build our anti-time cathedral?

Fortunately, yes. Although many casually use entropy as an argument for time, Boltzmann himself didn't see it that way. Entropy, he argued, is simply the result of living in a world of mechanically colliding particles where disordered states *are the most probable.* Because there are so many more possible disordered states than ordered ones, the state of maximum disorder is simply the most likely to appear. Put another way, entropy is merely a matter of things slamming into other things in the here and now. No arrows exist. Randomization is a present-moment process. Sure, we humans can always peer at a dynamic scene, look away for a while, then look again, and things will be different. But different scenes, the fact of change, and randomization itself are not the same thing as time.

Boltzmann essentially said that a state of order in which molecules just happen to all move at the same speed and direction is the most improbable case we can imagine. In other words, the second law of thermodynamics is merely a statistical fact. Any gradual disordering of energy is like shuffling a deck of cards. What we called "order" when the deck was purchased, with each suit arranged in ascending array, was a special case. The act of randomization requires no ghostly magical external entity.

So if time does not actually exist, what do we experience in everyday life? We need to know before tackling the ultimate scary time-consequence, the apparent end to life. But more importantly, we need to know who experiences what, where these adventures take place, and how our lives unfold.

QUANTUM GUYS WRECK THE POOL TABLE

5

> "Contrariwise," continued Tweedledee,
> "if it was so, it might be; and if it were so, it would be;
> but as it isn't, it ain't. That's logic."
> —Lewis Carroll, *Through the Looking-Glass,*
> *and What Alice Found There* (1871)

Most people believe that there's an independent physical universe "out there" that has nothing to do with our awareness of it. This seeming truth persisted without much dissent until the birth of quantum mechanics. Only then did a credible science voice appear, which resonated with those who claimed that the universe does not seem to exist without a perceiver of that universe.

Until then, this whole business was deemed a murky issue more appropriate to philosophy than to science. Yet the relationship between the physical world and consciousness, so redolent with the subjective aromas of cultural norms, has actually vexed and fascinated science for centuries.

On the face of it, consciousness or perception seems wholly different from the atoms, forces, and cause-and-effect machinations of the cosmos. If today one tried to unite them all, one's initial tendency would be to give primacy to the material universe and then to try to find a way in which consciousness sprang from it. For example, the brain is made of atoms, which are made of subatomic particles—all known entities—and it operates by an electrochemical process whose nature is no longer mysterious. If our awareness is merely some sort of subjectively felt spin-off of all this, then it could indeed be incidental and secondary to the modern world's self-operating model of reality, in which case you wasted your money purchasing this book. Science would have gotten away with exactly that model, had it not been for a little niggling matter that arose just over a century ago: quantum mechanics.

Basically—and this goes back more than two millennia to the days of Aristotle—an early issue was whether consciousness fundamentally belongs to a realm separate from the physical world. It wasn't a preposterous idea. Believing so allowed those who wanted to explore things like free will, morality, spirituality, and (later) psychology to have one arena to themselves, whereas those dealing with the hows and whys of the physical cosmos had another. The two didn't need to muddy the same waters.

If there was any connection or commonality between the two realms—of consciousness and the physical world—it was that the gods or the one God was universally assumed to have created both. This is why treatises on individual behavior, as well as the discoveries by "Natural Philosophers" like Newton, who successfully uncovered the logic and consistency for all physical motion, routinely cited the Creator. The practice only vanished during the past century. These days, neither your therapist nor your physics teacher is likely to bring up the Deity.

Even as late as the seventeenth century, René Descartes declared that two totally different realms inhabited the cosmos:

mind and matter. He had his own good logic for saying so, because in order for mind and matter to interact, there must be an energy exchange. And no one had ever observed any object's energy either shrink or grow simply because it was being observed. Naturally, if our minds do not affect matter, the reverse must also be true. And if the universe's total energy never changes (which is true), then it seems to leave no room for one or more separate consciousnesses to have any energy at all, which implies that consciousness doesn't even exist.

But it does, as Descartes illustrated with his most famous maxim. So from that point forward, scientists pretty much left consciousness alone. When halfhearted efforts to unite everything occasionally arose, they were always based on the primacy of the random and inert material world that presumably gave birth to awareness somehow. (This was sometimes called physical monism.) No one tried traveling the obverse route by attempting to argue that the material universe might arise from consciousness. This absence couldn't be faulted. Consciousness was and still is perceived as almost ghostly— how could mere perception move a rock, let alone create a planet?

Thus the choice was clear among thinking people. The verdict in modern science was, and still is, stick with the Cartesian dualism of mind and matter. For centuries they've been regarded as inherently separate—or, in the view of a growing majority, consciousness somehow arises from an as-yet-undiscovered mechanism within material bodies, such as the structure or chemistry of the brain.

The motive behind asserting a duality between mind and matter was both noble and logical. Aristotle, desperately wanting to figure out how things work and desiring to uncover the physical rules of the cosmos, felt that removing the error-prone opinions of individual observers could only improve things. In short, he fought for *objectivity*. This essentially maintains that everything in the world is separate and independent from our minds. Isaac Newton very much liked this idea, too, and by the middle of the seventeenth century, his three laws of motion helped cement what we now call classical physics.

In France at around the same time, René Descartes was fully on board with this assumption of *material realism*, or *causal determinism*. (Those fancy terms merely refer to our standard model of the universe as provided by Newtonian physics. It's simply the idea that all objects have mass and influence upon each other. Without the "pull" of all these myriad moving objects, everything else would remain at rest, or else continue traveling undisturbed, and we'd see no changes unfolding.) Remembering the harrowing travails of the likes of Galileo just a few decades earlier, Descartes figured that this assumption of material realism would let science proceed with the greatest safety and minimal interference from the Church. Let the Church have that other realm—of mind, consciousness, individual spirit, morality, societal rules, religious rituals, and whatever else they wanted—when it came to regulating personal behavior.

It worked. Science and the Church now had their own fiefdoms. The Newtonian–Cartesian view was that the cosmos is essentially a giant machine. Originally scientists paid a bit of lip service to the Deity, but essentially they viewed the universe as a giant, self-sustaining, three-dimensional game of billiards. If you knew the masses and speeds of each object, you could perfectly predict future positions and behavior, or even extrapolate in reverse and know where everything had been.

Similarly, in the next century, French mathematician Pierre-Simon Laplace surmised that if someone had sufficient intelligence and information, they could know everything about the universe just by observing the current positions and trajectories of all objects. Everything was determined by previous conditions. No mystery remained except, perhaps, for the small matter of ultimate origins. Not even God was necessary; indeed,

Laplace omitted any mention of a deity in his writings on celestial mechanics.[3]

Such was the view of reality in the closing moments of the nineteenth century and early years of the twentieth. Each side pretty much kept its bargain. Science left religion alone and ignored consciousness as well. And religion considered science to be okay—after all, it explained how things moved and didn't trespass into trying to figure out why or how the cosmos came to be.

As the Western world gained in living standards and concomitantly grew less religious, the scientific deterministic model became the new gospel. It was often called *scientific realism*, and who could argue with such a label? You'd have to be a nutcase to be antiscience or antirealism.

In sum, the universe was widely regarded as objective (existing independent of the observer), made of matter (which included energy and fields), ruled by causal determinism, and limited by locality. When it was even considered at all, consciousness

---

[3] It didn't have to be stated, but another element to this classical physics model was what Einstein later called *locality*. Nothing budges unless acted upon by a nearby object or force. Einstein famously showed that the ultimate speed, that of light at 186,282.4 miles per second, imposes a limit of how quickly anything could affect anything else.

Einstein explained that nothing with any mass (i.e., that weighs anything) can quite attain lightspeed, because its mass would grow until, for instance, even a feather at just below lightspeed would outweigh a galaxy. And the amount of force needed to accelerate such a huge mass further would be impossible to obtain—it would exceed all the energy in the universe. Indeed, at the speed of light, a zooming mustard seed would outweigh the entire cosmos. (This change of "weight" that automatically accompanies speed was part of Einstein's first, special relativity theory of 1905. It happens because motion always involves energy, and energy and mass, he said, are two sides of the same coin. They're equivalent, as per his famous $E = mc^2$, where the E is energy and the m is the object's mass. So if you increase an object's inherent energy by increasing its speed, you're also increasing its equivalent mass.) See chapter 7 for a wider discussion on the implications of locality.

or the observer was assumed merely to be part of the physical matter-based cosmos, having somehow arisen from it. That its origins or actual nature couldn't be explained seemed to bother no one. A few lingering mysteries were deemed perfectly compatible with the material universe.

And this is where we'd still be if it weren't for quantum mechanics.

That new branch of physics started quietly enough. Not much couldn't be explained by classical physics until the closing years of the nineteenth century, but puzzles were starting to grow. Some were just plain odd. For example, a bonfire and the Sun were both deemed to be blazing fires. (The Sun's true energy-releasing process of nuclear fusion wasn't explained until Arthur Eddington did so in 1920.) If you stood too close to a bonfire while holding out a hot dog or a marshmallow on a stick, you'd jump back because your skin could grow painfully hot—certainly more uncomfortable than solar rays ever make you feel, even at midday. And yet despite the ample heat, a bonfire can never deliver a tan or "sunburn." But why? This was unexplainable.

We'd known about ultraviolet (UV) rays since their discovery by Johann Ritter in 1801, and that such UV photons (bits of light) coming from the Sun are what produce suntans and sunburns. But why didn't we ever get any from a campfire? Classical physics said that UV should be present, and hanging out long enough around a campfire should deliver a tan. But it never did.

The answer had to do with electrons, which were discovered in 1897. They were immediately assumed to orbit around an atom's nucleus like planets around the Sun. But here's the thing: In 1900 Max Planck surmised that electrons can absorb energy from a hot environment, and then radiate it back in the form of bits of light, which ought to include some ultraviolet light. But if electrons—unlike planets, which can orbit the Sun at any distance at all—could only orbit their atom at specific, discrete locations, then they would only be able to absorb or emit specific quantities of energy, called *quanta* because it takes a precise amount or *quantum* of energy to move an electron a specific distance. If the environment wasn't energetic enough, electrons would only be able to make easy jumps, like those in the atom's outer fringes. They'd never be able to make a powerful jump from the innermost orbit to the next highest, which is what's required to create a UV photon when the electron fell back down again.

Planck's idea, soon called the *Planck postulate*, was that electromagnetic energy could be emitted only in specific quanta. It wasn't long before Niels Bohr, the brilliant Danish physicist, confirmed that all atoms indeed behave like that. Only by falling back inward from one allowable, higher orbit to another one closer to the nucleus do atoms emit packets of light, called *photons*. This is the only way in which light is born. If an atom is not stimulated, its electrons remain in stable orbits, and it produces no light at all.

That high-energy drop from the second orbit to the innermost one—needed to create a sunburn-producing UV photon—requires a more powerful initial energy boost than a campfire can provide. Quantum theory—the idea that electrons can make only specific moves between allowable orbits and thus absorb or emit only specific quanta of energy—explained previously enigmatic facets of nature. So far, so good. But weirdness was already lurking in the closet. According to Bohr, an electron cannot exist in any intermediary position outside a precise, allowable orbit; anytime it changes position it must go from one specific orbit to another, and never be anywhere between them. So here's what's odd: As an electron changes orbits, it does not pass through the intervening space!

Imagine if the Moon behaved like that. It used to be much closer to us, and is still moving farther away at the rate of almost two inches a year. It's spiraling away like a bent skyrocket. Also, physics allows the Moon to be any distance from us. Now imagine if the Moon didn't budge in its separation from us for millions of years, but then, in an instant, suddenly vanished and rematerialized in a new location fifty thousand miles farther away. And imagine, too, that it accomplished that jump in zero time without passing through any of the intervening space.

Well, that's what electrons do. Needless to say, this opened bizarre new implications and set the stage for earthquakes that rocked classical physics forever. Even Planck unsuccessfully struggled to understand the meaning of energy quanta. "My unavailing attempts to somehow reintegrate the action quantum into classical theory . . . caused me much trouble," he wrote with exasperation many years later. Ultimately he gave up trying to make logical sense of it, or even trying to convince his most stubborn doubters. "A new scientific truth does not triumph by convincing its opponents and making them see the light," he said presciently, "but rather because its opponents eventually die, and a new generation grows up that is familiar with it."

But it was hard for anyone to get too familiar with quantum mechanics because strange new revelations kept arriving. Physicists learned that light, as well as bits of matter, are not just particles but also are waves, and how they exist depends on who's asking—meaning, the method of observation determines how these objects appear! Actually it's worse than that. These entities can also exist in two or more places at once, in a kind of blurry probabilistic fashion. We might say that electrons acting as waves are really wave *packets*, and where the packet is densest is where an individual electron is most likely to materialize as a particle. But it may also, upon observation, pop into existence in an unlikely place, on the almost totally empty fringes of that packet. Over time, a series of observations will show electrons or bits of light materializing according to probability laws.

This means the electron or photon doesn't enjoy any independent existence as an actual object in a real place, with a real motion. Instead, it exists only probabilistically. Which is to say it doesn't exist at all—until it's observed. And who observes it? We do. With our consciousness.

Suddenly, consciousness and the cosmos—which had parted paths way back with Aristotle, and whose divorce seemingly was made more permanent by Cartesian and Newtonian credos—might not be such totally separate entities after all.

Slowly, in the opening decades of the twentieth century, classical physics and the common-sense gospel of locality were eroding. After all, some "motion" unfolded without the object penetrating through any space or requiring the slightest bit of time.

Objectivity was melting, too, because the observer alone made these tiny objects materialize. Causal determinism was vanishing as well, because nothing palpable or visible caused the entities to assume one position instead of another. And as for the "physical monism" that made consciousness a random offspring of the material cosmos, it now gained interest and was reexamined. It suddenly seemed like consciousness might enjoy some central importance in the universe's overall reality. After all, the observer's awareness was now seen to determine what physically occurs.

And yet despite these profound oddities being increasingly perceived in the 1920s, the real quantum strangeness was just beginning.

THE END OF TIME

6

Stand still, you ever-moving spheres of Heaven,
That time may cease, and midnight never come.
—Christopher Marlowe, *The Tragical History of the
Life and Death of Doctor Faustus* (1604)

When we grow up watching our loved ones age and die, we assume that an external entity called *time* is responsible for the crime. But as we've seen, many lines of science and logic cast doubt on the existence of time as we know it. We must repeat that, although we do observe change, change isn't the same thing as time.

So what are we experiencing? To observe change, such as motion from one point to another, we should examine the process—what actually is unfolding. Now, to measure anything's position precisely is to "lock in" on one static frame of its motion, like a single frame or screenshot of a film. Conversely, as soon as we observe movement, we can't isolate a frame, because *motion is the summation of many frames.* Sharpness in one parameter induces blurriness in the other.

Let's pay homage to Zeno and consider a film of his flying arrow. We can stop the projector on a single frame. The pause enables us to know the position of the arrow with great accuracy—there it is, hovering eight feet above the archery tournament field. But we've lost all information about its momentum. It's going nowhere; its path is uncertain.

What's interesting is that, since the 1920s, numerous experiments confirm that such uncertainty is not merely a matter of having insufficiently precise technology. Rather, uncertainty is built into the fabric of reality. This basic fact of nature was first expressed mathematically by German physicist Werner Heisenberg and today is of course universally known as Heisenberg's *uncertainty principle*.

The truth of this started to become clear when scientists measured objects like electrons. Increasing accuracy in figuring out their direction and speed (momentum) yielded ever-growing blurriness in knowing where they were at any given instant (position). At first everyone assumed that we'd eventually be able to nail both down with high certainty. In other words, our inabilities were due to our own technological immaturity, and we'd soon do better. We never did. Thus, an amazing thing soon became obvious. An electron doesn't *have* an exact position and an exact motion. Rather, the act of observing results in perceiving one characteristic or the other or else a vague sense of both. The uncertainty principle became a fundamental concept of quantum physics.

It may seem spooky, but the weirdness completely goes away, and the whole thing makes sense, when viewed from a life-based perspective. According to biocentrism, time is the inner sense that animates the still frames of the spatial world. Remember, we can't see through the bone surrounding the brain; everything we experience right now, even our bodies, is a whirl of information occurring in our minds. Space and time are merely the mind's tools for effortlessly putting everything together.

So what's real? If the next image is different from the last, then it's different, period. We can award change with the word *time*, but that doesn't mean that there's an invisible matrix in which changes occur.

We view life while perched on the edge of that paradox described by Zeno. Because an object can't occupy two places simultaneously, we can synopsize his conclusions by noting that an arrow is somewhere (and nowhere else) during each instant of its flight. To be in one place, however, is to be at rest, however momentarily. The arrow must therefore be motionless at each discrete moment. Thus, motion is not what's happening, at least not if we insist it's a time-based phenomenon.

Okay, it may be confusing to deny motion without elaboration. What we're really saying is that motion isn't a feature of the outer, spatial world, but rather a conception of thought. Evidence for this is provided by the fact that the observer affects the motion in the "external" world. An experiment published in 1990, which has actually been called the "quantum Zeno effect," shows that, according to physicist Peter Coveney, "the act of looking at an atom prevents it from changing." (In the next few chapters, we'll see how this actually works in the visible world.) Because space and time are forms of animal intuition, they're tools of the mind and thus don't exist as external objects independent of life. When we feel poignantly that time has elapsed, as when loved ones die, it constitutes the human perceptions of the passage and existence of time. Our babies turn into adults. We age. That, to us, is time. It belongs with us.

New experiments since 2000, explored in chapter 8, confirm this. These suggest that the "past"—the history of the cosmos, of Earth, or anything else—is not some fixed statue, but unfolds in the present moment *and only upon observation.*

Indeed, quantum mechanics insists that when it comes to the 10^{80} subatomic objects that comprise the observable universe, none have real existence or actual motion. The only things that are real, insists quantum theory, are *observed* events that emerge from the blurry possibilities that always exist.

This is so important, that we need to strap on our seat belts and really understand the experiments that changed time and space forever.

What follows now is science, not speculation. It so unequivocally supports the new worldview that it is vital in letting us see why biocentrism is not philosophy or speculation, but rather is rooted in observation and experimentation. The physics that follows is not difficult, and we've avoided equations and the most technical aspects. Nonetheless, those who are truly science averse, or perhaps don't care to understand how quantum mechanics supports the unity of nature and the observer, should feel free to skip ahead to chapter 9.

Quantum theory (QT) started, as we've seen, with the realization that in the land of the tiny—the realm that ultimately dictates what happens in our visible macrocosmic reality—particles do not behave as logic would demand. Very soon, proponents of QT learned that to be useful, to predict behavior in the physical universe, QT had to deal exclusively with probabilities. Thus the concepts of the *likely* places particles may appear, and *likely* actions they will take (as opposed to their definite locations and actions), became mainstream to physics. This was helpful and fine when it came to understanding nature. It wasn't too upsetting to grasp that on some levels, the best we could do was learn the *probability* of things occurring.

The truly strange aspects of QT really started rolling with its now-famous double-slit experiment. It is a new version of this experiment that we will focus on shortly. But first, in case you didn't read our first book (or even if you have, and could use a little refresher), here are the rudiments of this classic demonstration, first performed over a century ago and repeated countless times. It was this experiment that first showed without a doubt that the observer intimately influences what is perceived.

It began when scientists were still trying to figure out the nature of light. Isaac Newton had insisted that light is made of particles, but other investigators soon seriously doubted this was true. In the early nineteenth century, British scientist Thomas Young, by passing a ray of light through variably spaced holes, showed that this arrangement produced an odd series of bands. This proved that light consists of waves, since the pattern was consistent with an alternating series of subtracting and enforcing interferences, which only waves would produce. (Bullets or particles can never erase each other, whereas the peak of one wave, when meeting the trough of another, will cause both to cancel out and disappear entirely.)

For nearly a century thereafter, physics flat-out decreed that light consists of waves. But the 1887 observation of a curious phenomenon that soon became known as the photoelectric effect—the 1905 explanation of which won Einstein his Nobel Prize—revealed that under different conditions, light acts as if it's made of a series of discrete, massless bullets. Einstein's explanation of this wave–particle dichotomy was actually one of the early impetuses for quantum mechanics.

The first modern double-slit experiment was performed in 1909 by British physicist Geoffrey Taylor. It starts by aiming light at a detector wall. (These days the experiment can use "solid" subatomic particles like electrons or instead use light, but back then only light was practicable.) Before hitting the wall, however, the light must pass through an initial barrier with two holes (referred to as right and left slits) in it. Each bit of light has a 50/50 chance of going through the right or the left slit.

We can shoot a flood of light or just one indivisible photon at a time, and the results remain the same. After a while, all these photon-bullets should logically create a pattern—falling preferentially behind each slit, since most paths from the light source go more or less straight ahead. Logic says that we should see a cluster of hits behind each opening, as is shown in Figure 6-1:

Figure 6-1. Photons or electrons fly through the slits and should logically create detectable "hits" behind each opening.

But that's not what happens. Instead, we get a strange pattern that looks like Figure 6-2:

Figure 6-2. In actuality, an interference pattern materializes, indicating the presence of interacting waves. This pattern is reliably seen even if only one photon or electron at a time is allowed to pass through the openings. But how is this possible? With what is this lone photon or electron interfering?

Turns out, this pattern is exactly what we'd expect if light is made of waves, not particles. Waves collide and interfere with each other, causing ripples. If you toss two pebbles into a pond at the same time, some waves meet each other to amplify the height of the new colliding wave, or one wave might encounter the other's trough, in which case these cancel out and the water is flat in that spot.

So this early-twentieth-century result of an interference pattern, which can only be caused by waves, showed physicists that light is a wave, or at least it acts that way when this experiment is performed. The fascinating thing is that when solid physical bodies like electrons were later used, they got exactly the same result. Solid particles have a wave nature, too! So, right from the get-go, the double-slit experiment yielded amazing information about the nature of reality. Unfortunately, or fortunately, this was just the appetizer. Few realized that true strangeness was about to be served steaming hot.

The first oddity happens when just one photon or electron is allowed to fly through the apparatus at a time. After enough have gone through, are individually detected, and start to build up a pattern, this same interference arrangement emerges. But how can this be? *With what* is each of those electrons or photons interfering? How can we get an interference pattern when there's only one indivisible object in there at a time?

There has never been a satisfactory answer for this that employs simple logic or classical physics. At first, wild ideas kept emerging. Could there be other electrons or photons "next door" in a parallel universe, from another experimenter doing the same thing? Could their electrons be interfering with ours? That's so far-fetched that few believed it.

The usual interpretations of why we see an interference pattern—now accepted pretty universally—is that photons or electrons have those two choices of slits when they encounter the double holes, but do not actually exist as real entities in real places until they are observed, and they aren't observed until they hit the final detection barrier. So when they reach the slits, they exercise their probabilistic freedom of taking *both* choices. Even though *actual* electrons or photons are indivisible and never split themselves under any conditions whatsoever, they do not actually become electrons or photons until they are observed—and they reach those slits before they are observed.

Thus they exist as pre-photon or pre-electron "probability waves," and different rules apply. What goes through the slit are not actual entities but just ghostly probabilities. Each probability wave of each individual photon interferes with itself! When enough have gone through, we see the overall interference pattern as all probabilities congeal into actual entities making impacts and being observed—as waves. A probability wave (which no one can really visualize) can be imagined as a precursor or tendency toward the actual existence of a photon or electron, which never achieves any reality as such entities unless observed. It's as if it doesn't exist, yet at the same time exists as all possibilities.

Sure it's weird, but this, apparently, is how reality works. And this is just the very beginning of quantum weirdness. QT has a principle called *complementarity*, which says that we can observe objects to be one thing or another—or to have one position or property or another—but never both. This again is linked with that famous Heisenberg uncertainty principle, which says that the more precisely we pin down one aspect of an object—like its position—the less well known its motion becomes. It depends on what one is looking for, and what measuring equipment is used. In reality, Heisenberg said, all possibilities simultaneously exist, until a single one materializes upon observation.

Suppose we wish to know which slit a given electron or photon has gone through, on its way to the barrier. It's a fair enough question, and it's easy enough to find out. We can use polarized light. This is light whose waves are not all scrambled up the way they usually are, but instead vibrate either horizontally or vertically. (Their orientation can also be slowly rotated, but let's keep this as simple as possible and leave such mutating "circular polarization" out of this discussion.) Light is polarized in nature when it is reflected, for example, which is why your sunglasses can remove the glare from windows or ocean surfaces—they have been treated to block the reflected, polarized light. Yet if you cock your head, suddenly those reflections appear. Thus, each polarized lens is set at an angle that allows only one of the two kinds of photons to pass through the gap, effectively tagging these bits of light and letting us know the which-way path the photon traveled.

When a mixture of assorted polarizations is used, we get the same result as before. But now let's determine which slit each individual photon is passing through by using light of either a "vertical" or a "horizontal" polarization. Many different techniques have been used, but it doesn't matter which method we choose. The important point is that we employ a setup that lets us determine the "which-way" information for each electron or photon as it heads through one of the gaps toward the detector.

So we repeat the experiment, shooting photons through the slits one at a time, except this time we will learn what slit each photon traverses. We can gain such which-way knowledge by placing polarizing lenses in front of each opening—as depicted in Figure 6-3—and shooting a scrambled ensemble of light containing photons with both horizontal and vertical alignments. The polarizing lenses act like markers or tollbooths. Each lens blocks all light except for photons with the correct polarization. So if we have a "vertical" polarizer on the right slit, then we know that only vertically polarized photons can penetrate it and strike the final barrier.

By having the lens in front of the right slit oriented to one polarization and the left slit guarded with the opposite polarization, we will learn which way each photon went, because only an "up/down"-oriented photon can penetrate the right lens (say) and only a "sideways" photon can go through the other one. In short, we have gained which-way information.

Astonishingly, the *results* now dramatically change. Even though our which-slit detector is known not to alter photons or electrons, we no longer get the interference pattern seen in Figure 6-2. Now the results suddenly change to what we'd expect if the photons were particles—a mass of "bullet" hits on the detector screen behind each slit, as in Figure 6-1. The wave pattern, showing interference, is gone.

Figure 6-3. Polarized lenses let observers determine which slit each photon passes through. This which-way knowledge in the scientists' minds somehow causes each bit of light to lose its freedom of simultaneously taking both paths, and forces it to materialize into an actual object (photon) *before* entering the slits. This in turn makes the interference pattern vanish. Instead, we now see simple hits behind each opening.

Something's happened. Turns out, the mere act of measurement, of learning the path of each photon, destroyed the photon's freedom to remain blurry and undefined and take both paths until it reached the final detection screen.

Its probabilistic *wave-function* must collapse at our which-way measuring device, because this time we're essentially observing (gaining knowledge) *before* the photon hits the slits as well as at the detector in the back. Its wave nature was lost the instant each photon lost its blurry, probabilistic, not-quite-real state. But why should the photon have chosen to collapse its probabilistic wave-function? How did it *know* or care that we, the observer, could learn which slit it went through?

Countless attempts to get around this, by the greatest minds of the past century, have failed. Our *knowledge* of the photon or electron's path alone caused it to become a definite entity ahead of the previous time. Of course physicists also wondered whether this bizarre behavior might be caused by some interaction between the which-way detector, or various other devices that have been tried, and the photon. But no. Totally different which-way detectors have been built, none of which in any way disturbs the photon. Yet we always lose the interference pattern, since the measured photon invariably changes its nature from a wave to a discrete particle. The bottom-line conclusion, reached after many years, is that it's simply not possible to gain which-way information *and* produce the interference pattern caused by energy waves.

This which-way experiment illustrates that photons can exist as a particle, which it must be if it is to pass through just one slit and not both, or as a wave, which blurrily penetrates both simultaneously. But they cannot be seen to be both a particle *and* a wave. Again, the main point is that *where* we observe the photon or electron is what makes it become one or the other. And just in case you're suspicious of the detectors, note that when used in all other contexts, including double-slit experiments without information-providing which-way readouts at the end, polarizing lenses never have the slightest effect on the creation of an interference pattern.

We're left with no choice but to accept that our presence as an observer, and how we make the observation, physically changes what we're looking at. But we need more persuasion. Turns out the tool for reaching the next level of proof arrived with one of quantum theory's wildest realities: particle entanglement.

A BIZARRE REALM OF ENTANGLED TWINS

7

The stars up there at night are closer than you think.
—Doug Dillon, *Sliding Beneath the Surface* (2010)

What's the strangest, most mysterious aspect of this amazing universe?

There's no shortage of candidates, but one really stands out. It seems baffling, although it's now universally accepted among physicists. Its exploration requires a quick peek into the intriguing realm that surrounds the speed of light, which until recently seemed like the universe's absolute speed limit.

In 1905, Einstein gave meaning to a wild observation made in the previous few decades by Hendrik Lorentz, George FitzGerald, and others. They had all realized that light travels at a constant speed, and they understood how profoundly remarkable this is. It means that photons from the headlights of a rapidly approaching car strike you at light's unwavering rate of 186,282.4 miles per second, the same as if the car weren't moving. Or consider: Earth's orbit propels us toward the star

Deneb in June, but we zoom away from it in December, yet its light acts as if we're stationary and forever hits us at the same speed. Imagine if wind acted like such a constant and felt like an unvarying gentle breeze regardless of whether you were stationary or holding your arm out of a fast-moving car or plane. So right from the get-go, light starts out strange and unique.

Of course, as we've seen, there was also the small matter of what, exactly, light is. In the last chapter we saw that physics ultimately found it could be a wave or a particle, depending on the observer and the experimental method. Later, in the present Standard Model of How Things Work, a photon came to be regarded as a force-carrying particle, like a butler, delivering the electromagnetic force from one place to another.

So what exactly is light? It's a straightforward question, but the answer is a bit murky, because bits of light (photons) act differently depending on the way we detect and analyze them. When encountering an object, a photon acts like a particle, sort of like a tiny bullet that has energy but actually weighs nothing (assuming you could bring it to a stop and weigh it, which you can't). Its power depends on its color. Violet photons are more energetic than red ones.

If a photon hits a bit of metal, it can knock loose electrons as if it were a bullet and in a way only a particle can achieve. But while on its merry way between destinations, it's probably better to visualize it as a wave of energy. Actually, two waves. Each bit of light is a magnetic "pulse" or "field" of energy that wavers in intensity in an on-and-off fashion. Traveling toward it at right angles is another wave or field—an electric field. Both waves are what make up a single photon. Each field creates the next, so this entire dual-wave entity is called an *electromagnetic* wave. This zooms along at its famous unvarying speed, fast enough to whiz more than eight full times around the Earth in a single second.

Recently, researchers made headlines by dramatically slowing light. We've always known that light automatically slows down when traversing air, water, or other dense media. Sunlight going through your window glass decelerates to about 120,000 miles per second, and then instantaneously speeds up again once it's through. In truly dense but translucent materials under certain conditions, light can be brought to a near halt. Photons have recently been slowed to 38 miles per hour. Imagine—light that wouldn't even get a ticket on the freeway!

Outside a vacuum, we can shoot particles through substances where they readily exceed light's velocity in those substances. As an example in nature, electrons jazzed by powerful magnetic fields near some massive stars (i.e., the synchrotron process) can be hurled through a nebula faster than the speed of light in that medium. This creates a beautiful blue shockwave called *Cherenkov radiation* as the particle breaks the "light barrier." Light is strange, but you can get used to it.

Through it all, light's sovereignty in a vacuum has never been seriously challenged. Until now. Rewind to our odd world of quantum mechanics. This realm, which makes Alice's adventures a comparatively tame walk in the park, is a Wonderland where we've seen that particles can simultaneously exist and not exist, only to spring into reality as soon as someone takes a look.

The idea of a kind of inseparability so far as space and time are concerned was dealt with by John Bell in the 1960s. His idea was that particles—whether of matter or of light, it doesn't make a difference—don't really independently exist except as a kind of probabilistic entity. (Can't picture that? You're not alone.) The act of observation causes this mere probabilistic wave-function to "collapse," and the object abruptly materializes as an actual entity in a real location. In short, the classic idea of an atom's nucleus being orbited by one or more electrons that each have an independent existence at every moment in an actual place and with actual motion must be discarded. Instead we should think of them as existing in a kind of blurry state called *superposition*, where virtually anything that could happen exists on some level, ready to materialize. At the moment an experiment or observation is done, the electron leaves this probabilistic existence and appears in physical reality.

With *entanglement*, two particles are born together, as when a light is shined into certain crystals like beta barium borate. Inside the crystal, an energetic violet photon from a laser is converted to two red photons, each with half the energy (twice the wavelength) of the original, so there's no net gain or loss of energy. Out pop the two photons, which fly off in different directions, but which secretly share a wave-function. If one is observed, its wave-function and that of its twin simultaneously collapse—regardless of the distance between them.

Even if the twins are separated by half the diameter of the universe, says quantum mechanics, observing one twin will cause both to become actual entities. Moreover, they must exhibit complementary characteristics. With light, a photon can have a horizontal or a vertical orientation (polarization) of its waves. With an electron, it might exhibit an "up" spin or a "down" spin. So when the twins' "potential" or "wave-function" collapses, both cease being blurry, not-really-there objects, and now suddenly materialize as actual entities. One photon or electron will have one aspect (e.g., it can spin upward or downward, or be polarized horizontally or vertically), while its twin always exhibits the other property—the complementary attribute.

When either is observed, the real shocker is that its twin "knows" what happened to its doppelgänger (i.e., that it came into an actual physical existence as a photon or electron) and instantly assumes the complementary guise—even if its twin is in a different galaxy. During this process no time will elapse, no matter their distance apart. It's as if there's no space between them. They're essentially two sides of the same coin, and distance between them is nonexistent, even if, to us, it's half the width of the cosmos.

Einstein hated this because he believed in locality—that an object can only be acted upon by something in its neighborhood. A leaf in Brooklyn would be stirred by a local gust of

wind, but it won't be instantaneously jostled by the air distur-
bances generated by a lively peasant revolt on an alien planet in
the Andromeda Galaxy.

In 1935, Einstein and two colleagues, Boris Podolsky
and Nathan Rosen, wrote a now-famous paper in which they
addressed this aspect of QT. Examining the prediction that
particles created together—entangled particles—can somehow
know what the other is doing, the physicists argued that any
such parallel behavior must be due to local effects, some con-
tamination of the experiment, rather than some sort of "spooky
action at a distance." The paper was so celebrated that such
synchronized quantum antics became known as "EPR correlations" after the initials of the surnames of those three physicists, and the "spooky action" line became endlessly quoted as
a pejorative, a put-down of this ridiculous idea that—on some
fundamental level—there could be no space between objects or
no time lapse between events.

A lot hinged on this. In a way, it was a pivotal time between
clinging to classical deterministic physics and accepting locality, as Einstein insisted on doing, versus traversing the strange,
blurry, quantum alleyways that, ironically, Einstein had helped
create with his 1905 explanation of the photoelectric effect.

Material realism (yet another label for the classical view-
point) says that physical objects are real regardless of whether
they are being observed. Moreover, unless they're in contact,
emit something like photons that can create contact, or at least
under some sort of influence via an electric, magnetic, or gravitational field, individual objects cannot influence each other.
And certainly they cannot do so if they are so widely separated
that electromagnetic energy from one does not have time to
reach the other.

As for instantaneous influence involving no time at all, or
influence that acts as if no intervening space exists between the
objects—forget about it, said Einstein and his colleagues. In
short, locality rules.

The opposite view at that time, taken by the likes of physicists such as Niels Bohr, then by Paul Dirac, and later by John Wheeler, is that objects can be *entangled*, or connected so that they are essentially inseparable. Observation of one object, or measuring it (which is the same thing), affects the other in real time. It doesn't matter how far apart they are. It's as if neither time nor space exist. Moreover, the "collapse of the wave-function" of one, so that the object goes from nonexistence or a kind of mere probability potential to being an actual object, has the same effect on the other—as if the observer and both objects were in the same place all together, at the same time.

No way, said Einstein. The whole house of cards—the objective universe, the independence of matter from consciousness, the belief in locality, the entire material realism of classical physics—hinged on this issue, and Einstein was not about to trade a cosmos ruled by logic and beautiful billiard-like machinery for one in which, as he put it, things would only materialize probabilistically. "God does not play dice," he famously sneered. Einstein could not accept that anything could just pop in and out of existence on the basis of mere likelihood or observation— especially if the observer wasn't touching the object in any way, but merely learning about it!

In other words, if these EPR correlations were what they seemed to be, then not only are the entangled objects not in any kind of contact with each other (which demolishes locality), but the observer whose awareness causes these events to unfold must be manifesting a consciousness that is also nonlocal, and indeed capable of "spooky action at a distance." As Erwin Schrödinger said in 1935, "It is rather discomforting that quantum theory should allow [a pair of objects] to be steered or piloted into one or another type of state at the experimenter's mercy in spite of the [observer] having no access to it."

It cannot be emphasized enough: In classical physics, objects or bits of light have definite properties, like existence in some location. Plus, they possess actual motion or spin with an axis pointing in some direction, or polarization—and the universe is filled with such objects that have these characterizations independent of our measurement or awareness of them. This, again, is what Einstein believed.

Quantum theory, by contrast, insists that nothing has location or momentum or spin or polarization unless it is measured. This is why the famous physicist John Wheeler said, "No phenomenon is a real phenomenon until it is an observed phenomenon."

Recent experiments (see chapter 9) have shown that Einstein was wrong. It is important for us to understand exactly how we know this, and how such demonstrations work, so we don't go away thinking that the issue is still up in the air in any way. It's also vital that we not bestow on QT incorrect powers or notions, such as those seen in some popular movies—like the idea that QT says that we can individually control the future, which is utterly untrue. Quantum theory is bizarre enough as is without giving it additional fictional attributes.

It should also be noted, as we discuss entangled twins instantaneously exchanging knowledge through space-time, that Einstein never visualized space-time as a kind of absolute physical gridwork, as if it were some sort of three-dimensional graph paper permeating space. Rather, he created the concept as a way to mathematically make sense of how observers in different reference frames (those moving at different speeds or experiencing differing gravitational fields relative to each other) would each perceive the passage of time or the lengths of objects or measured distances. Using his general relativity field equations, contradictions and paradoxes between observers were resolved. Those equations revealed how each observer would measure distance, mass, or time.

As a consequence, Einstein's equations also demoted space and time from being inviolable; no longer was the distance between an object and anything else absolute. No longer did a time interval have to be the same to observers in all places.

Instead, a single second could elapse for one observer while a thousand years simultaneously passed for another! Thus, despite the public's widespread, incorrect sense that time and space are actual entities, and this volume's several chapters disproving it, Einstein actually already disproved it more than a century ago.

And, just as a reminder, the removal of space and time as "constants" or absolute realities pulls the rug from under commonly held visualizations of reality, in which a physical universe dominated by objects floating in space was created at a specific inviolable time, and continues to abide within a temporally based framework.

THE MODERN
QUANTUM WORLD

8

"Spooky Action at a Distance"
—Iron Chic, song title (2013)

I n 1997, a Geneva researcher named Nicolas Gisin created pairs of entangled photons and sent them flying apart along optical fibers. When one encountered the researcher's mirrors and was forced to make a random choice to go one way or the other, its entangled twin, nearly seven miles away, always instantaneously acted in unison and invariably took the complementary option.

"Instantaneous" is the key word here. The reaction of the twin was not delayed by the amount of time that light would have traversed those seven miles—it happened at least ten thousand times faster, which was the experiment's testing limit. The echoed behavior was presumably simultaneous. Indeed, quantum theory predicts that an entangled particle knows what its twin is doing and instantly mimics its actions even if the twins live in separate galaxies billions of light-years apart.

This is so bizarre, with implications so enormous, it drove some physicists to a frantic search for loopholes. But in 2001, National Institute of Standards and Technology researcher David Wineland eliminated one of the main criticisms expressed by those who thought that the previous experiments failed to detect enough of the particle-events. (They had argued that this had introduced a bias by somehow letting observers preferentially see only those twosomes that acted in unison.)

Wineland used not light but solid, massive objects— beryllium ions—and his equipment had a very high detector efficiency. It was able to observe a large enough percentage of the in-sync events to seal the case. So this fantastic behavior is a fact. It's real. But how can a material object instantly dictate how another must act or exist when they are separated by large distances? Few physicists think that some previously unimagined interaction or force is responsible. Striving to understand, one of the authors personally asked Wineland what he thought, and he expressed an increasingly accepted conclusion: "There really IS some sort of spooky action at a distance." Of course, we all know that this clarifies nothing.

So there it is. Particles and photons—matter and energy— apparently transmit knowledge across the entire universe instantly. Light's travel time is no longer the limit. This is very big news because Einstein's relativity insisted—and all experiments for the past century have confirmed—that nothing can go faster than light. Anything with even the slightest bit of weight or mass, even a puff of incense smoke, could not be accelerated to lightspeed no matter the source of propulsion. And weightless entities like theorized gravity waves or photons of light can never exceed lightspeed. So this quantum knowledge, if that's what it is, where one object "responds" to the situation of another in zero time, meaning at infinitely fast speed, is stunning. Some physicists say that this does not violate relativity's light-velocity speed limit, the reason being that *we* cannot send information faster than light because the "sending" particle's information is governed by chance, and is thus not controllable. Others think that's a cop-out, that lightspeed limitations are now overturned and we've got to accept it and move on. In any case, we're walking a fine line about what constitutes information. *Something* is being conveyed instantaneously.

With this in mind, let's go back to the double-slit experiment of chapter 6, but this time we'll use entangled photons or entangled bits of solid matter. Remember, one commonly heard escape clause in the double-slit business is that our measuring devices bias the photons or electrons and change them, so that it's not merely knowledge in our minds that physically changes the results. But this objection has been dispensed repeatedly and effectively.

For example, in 2007, *Scientific American* reported an experiment where the which-way information (learning which slit each photon or electron goes through) employed polarized lenses positioned before each slit with their axes at right angles to each other. As we saw in chapter 6, this lets a beam of light containing mixed polarizations be separated out so that we'd know which slit each photon penetrated. As expected, the interference pattern was eliminated, as in Figure 8-1. Remember, as soon as we can learn which gap was traversed by each photon, all evidence of waves hitting the detector at the back vanish, and the pattern instead shows discrete separate hits.

Figure 8-1

But might it have been the polarizing filters that caused the wave nature of the light to vanish? Maybe filters do something to light, and it has nothing to do with ourselves as observers. No! Introducing a different polarizer in front of the detector with an axis of 45° relative to both slits "erases" all useful information about polarization, since now random photons get through both openings and we have no usable which-way information. The moment this "scrambling" filter was inserted, the interference pattern reappears, and now looks identical to what we see when there's no which-way measurements at all, as in Figure 8-2.

Figure 8-2

Experiments using light are a bit easier. Actual "solid" particles offer more of a challenge—especially when only one object at a time is allowed to pass through the apparatus. The first single-electron experiment to use an actual double slit wasn't reported until 2008 by Giulio Pozzi and his colleagues. The Italian team also conducted the experiment with one slit plugged, which—as expected—did not lead to the creation of a double-slit diffraction pattern. It took until 2012 before the team could perform an electron experiment in which the arrivals of individual electrons from a double slit were recorded one at a time. The point is just that science has now fully confirmed all these double-slit conclusions using not just bits of light, but also bits of matter.

Still, the most astounding double-slit experiments didn't start enchanting the world until the end of the twentieth century, when particle entanglement started to be utilized. So now let's use a device that shoots off entangled twins in different directions, using the barium borate crystal, a generator of entangled photons. Experimenters send these entangled photons off in separate directions. We'll call their paths' directions A and B.

We'll set up our original experiment, the one where which-way information is measured by using polarization filters, except now we add a "coincidence counter." The coincidence counter serves a single purpose: Switching it on or off either permits or prevents us from learning information, while it completely stands apart from the photons traveling through the double-slit apparatus. The way it works is simple. Its circuitry blocks all information about the polarization of each photon at detector A—and thus its "which slit" information—unless its entangled twin photon *also* hits detector B at around the same time.

To review, we've consistently seen that the moment we can learn which path each photon takes—and the polarization filters let us do this because each lens only allows the passage of either horizontal or vertical light waves—the pattern on the final detector abruptly changes, revealing that the photons have changed from waves to particles.

In the coincidence counter version of this experiment, the twin photons (A and B) follow separate routes to two detectors (A and B), but there is only one double-slit apparatus, in the path of photon A; photon B travels directly to detector B. Only when both detectors register hits at about the same time do we know that both twins have completed their journeys. The coincidence counter notes that the two photons have both been detected, and only then does something register on our equipment (Figure 8-3).

If we run this without any path-measuring polarization lenses in place, the resulting pattern at detector A is our familiar interference pattern, Figure 8-2. This makes sense. We haven't learned which slit any particular photon has taken, so they have remained probability waves until they hit the final screen.

Now we'll restore the polarization lenses in front of each slit, supplying which-way information for photons traveling along path A. As expected, the interference pattern instantly vanishes, replaced with the particle pattern, as in Figure 8-1.

So far so good. But now let's get tricky. Let's destroy our ability to learn the which-way paths of the A photons *without physically interfering with them in any way*. We'll even leave the polarization lenses in place. We'll merely switch off the coincidence counter. Because a coincidence counter is essential here in delivering information about the completion of the twins' journeys, figuring out anything about those paths has now been rendered impossible. The entire apparatus will now be useless for enabling us to learn which slit individual photons take when they travel along path A, because we won't be able to compare them with their twins—since nothing registers unless the coincidence counter allows it to. And let's be clear: We've left the slit-registering devices in place for photon A. All we've done is to remove our ability to gain which-way knowledge. (The setup, to review, delivers information to us—registers "hits"—only when the A photons register at detector A *and* the coincidence counter tells us that their twins' completed journey has been simultaneously registered at detector B. Shut off the coincidence counter, and no information registers.)

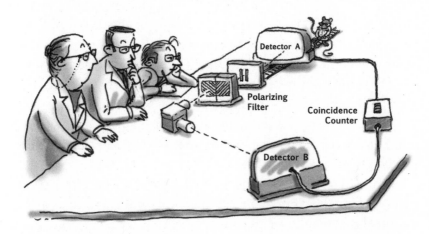

Figure 8-3. Adding a coincidence counter allows us to either gain knowledge of the experimental results, or else shuts off our data before we can learn anything—without meddling with the rest of the apparatus in any way. The movable distance to detector A (top) allows a further inquiry: By reducing the distance to the detector, and thus the time required for the A photons to reach it, we can learn what happens when the B photons complete the journey to their own detector (bottom) *after* the A photons have finished their own trip. The results indicate that time has no reality in the quantum world.

The result: They're waves again. The interference pattern is back. The physical places on the detector where the photons taking path A hit have now changed. Yet we did nothing to these photons' paths, from their creation at the generator all the way to the detector. We even left the slit-measuring devices in place. All we did was meddle with our ability to learn information via the coincidence counter. The only change was in our minds.

How could photons taking route A possibly know that we switched off some equipment somewhere else, far from their own paths? QT tells us that we'd get this same result even if we placed the information ruiner (the turned-off coincidence counter) at the other end of the universe.

By the way, this also critically proves that it wasn't those slit-measuring devices, the polarizing filters in and of themselves, that were causing the photons to change from waves to particles, thus altering the impact points on the A detector. We now get an interference pattern even with them in place (but when the coincidence counter is switched off). It's our knowledge that the photons or electrons seem concerned about. This alone influences their actions.

This is bizarre. Yet these results happen every time, without fail. They're telling us that an observer's mind determines physical behavior of external objects.

Could it *get* any weirder? Hold on—thus far the experiment has involved erasing the which-way information by turning off the coincidence counter. Now we'll try something even more radical—an experiment first performed in 2002. First we'll put detector A on a track so we can reduce the distance the A photons travel before they're detected, thus taking them less time to get there. This way, photons taking the B route will hit their own detectors *after* the A photons have finished their journeys. (The coincidence counter is turned on, so data is flowing.)

But oddly enough, the results do not change. When we insert the which-way lenses into path A, the interference pattern is gone, even though the coincidence-measuring ability that lets us determine which-way info for the A photons *will not occur until later.* But how can this be? Photons taking the A path already completed their journeys. They either went through one or the other slit or both. They either collapsed their wave function and became a particle or they didn't. The game's over, the action's finished. They've each already hit the final barrier and were detected—*before* twin B finished its own journey and thus triggered the coincidence counter into delivering useful which-way information.

The photons somehow know whether or not we will gain the which-way information *in the future*. Somehow, photon A knows whether the which-way data will *eventually* be present. It knows when its interference behavior can be present, when it can safely ride through both slits while remaining in its fuzzy both-slits ghost reality, or when it can't—because it apparently knows whether photon B, far off in the distance, will or will not *eventually* hit its detector and activate the coincidence counter that will ultimately deliver a useful signal to us.

It doesn't matter how we set up the experiment. Our mind and its knowledge or lack of it is *the only thing* that determines how these bits of light or matter behave.

These results are in keeping with what acclaimed physicist John Wheeler said as far back as the 1970s. Grasping the significance of John Bell's mathematical work involving wave-function collapses, he decided that only the observer determines the reality—it does not exist otherwise. Wheeler's 1978 article "The 'Past' and the 'Delayed-Choice' Double-Slit Experiment" was the inspiration for these just-described experiments a quarter century later.

As Wheeler explained at the time, "Nature at the quantum level is not a machine that goes its inexorable way. Instead what answer we get depends on the question we put, the experiment we arrange, the registering device we choose. We are inescapably involved in bringing about that which appears to be happening."

Posing an example, he set up a fascinating mind-experiment. Utilizing the fact that a strong mass or gravity warps space-time, he imagined a small, distant light source like a quasar, whose bits of light must traverse the vicinity of a foreground massive galaxy en route to our eyes. If the geometry is correct—if the distant quasar, the intermediate massive galaxy, and our Earth are all on a perfectly straight line—each photon's path will be warped to pass either above or below that galaxy. (The photon cannot go straight through the foreground galaxy because the galaxy's mass has altered the actual geometry of space-time so that the shortest "highway" from the quasar to Earth is no longer a seemingly straight line. In any case, the material in the foreground galaxy would block the quasar's light from penetrating it, even if it tried to travel that way.) Then it will continue for billions of more years before reaching our telescopes here on Earth (see Figure 8-4).

If they really had a 50/50 chance of taking either route, which path did each photon traverse? Wheeler's conclusion: The event, billions of years ago, *didn't really happen until we observe it today.* Only now will a particular photon pass above or below the foreground galaxy billions of years ago. In other words, the past isn't something that has already irrevocably occurred. Rather, long-ago events depend on the present observer. Until they're observed at this moment, the events didn't really unfold, but lurked in a blurry probabilistic state, all ready to become an actual "past" occurrence only upon our current observation. This astonishing possibility is called *retrocausality.*

Seems impossible, but experiments looking at the wave versus particle natures of distant quasar light are actually under way, with supportive results so far.

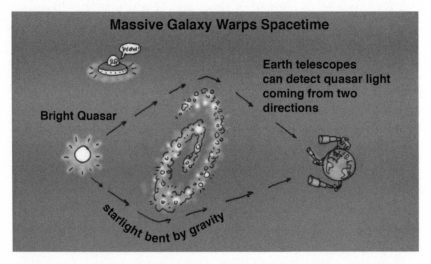

Figure 8-4. Our observations right now determine which path through space was taken by a photon from a distant quasar billions of years ago.

Although *retrocausality* is still under investigation, the instantaneous nature of quantum events is no longer in doubt. Moreover, although some think that this behavior is limited to the quantum world, the "two-world" view (i.e., the view that there is one set of laws for quantum objects and another for the rest of the universe, including us) is being investigated in labs around the world. In 2011, researchers published a study in *Nature* suggesting that quantum behavior extends into the everyday realm. Indeed, QT itself says that the effects should fully extend into our macroscopic everyday world.

In the October 2010 issue of *Scientific American*, theoretical physicists Stephen Hawking and Leonard Mlodinow stated, "There is no way to remove the observer—us—from our perceptions of the world . . . In classical physics, the past is assumed to exist as a definite series of events, but according to quantum physics the past, like the future, is indefinite and exists only as a spectrum of possibilities."

It's amazing that this breakaway from classical physics is still relatively unknown by the public, even if most people do equate quantum theory with strangeness.

How can we picture what's going on? If one physical object—an atom, a photon, or even a molecule—can "collapse" from a mere probability to an actual object, and simultaneously its twin knows this and assumes mirror-image properties, even if it's on the opposite side of the universe, how can we mentally picture the operating mechanisms? Perhaps it's best to assume that another realm pervades reality—a realm outside the space-time wherein planets orbit stars.

Thus when Einstein scoffingly sneered, "spooky action at a distance," he wasn't overreacting to what seemed to be unfolding. This realm is indeed spooky. Physicists refer to any communication between distant objects, employing no signals, as "nonlocal correlation." Others simply say it means that the twins represent two sides of the same coin—as if this provides any explanation!

What it really means is that there is an underlying reality that connects all the universe's contents. In this place, no separations exist between anything and anything else. Yet this realm creates events that materialize in space-time, in the observable physical cosmos.

To rephrase, in classical physics you cannot have instantaneous connections between objects—not in the universe in which we've always imagined ourselves to live. The distance between, say, Earth and Saturn requires more than an hour for light to travel, or a few years for our best spacecraft. It's a genuine separation. Yet at the same time, this space is part and parcel of a unitary system in which objects on Earth and on Saturn are in simultaneous contact.

Experiment after experiment continues to suggest that we—consciousness, the mind—create space and time, not the other way around. Without consciousness, space and time are nothing. This consciousness is co-relative with objects in that space-time realm. The conclusion seems inescapable. Suffusing the cosmos is the realm of mind, whose observations cause objects to materialize, to assume one property or another, or to jump from one position to another without passing through any intervening space.

These results have been described as beyond logical comprehension. But these are real experiments that have been carried out so many times that no physicist questions them. As Nobel-winning physicist Richard Feynman once remarked, "I think it is safe to say that no one understands quantum mechanics . . . Do not keep saying to yourself, if you can possibly avoid it, 'But how can it be like that?' because you will go 'down the drain' into a blind alley from which nobody has yet escaped."

But biocentrism makes sense of it all for the first time, because the mind is not secondary to a material universe. Rather, it is one with it. We are more than our individual bodies, eternal even when we die. This is the indispensable prelude to immortality.

NOTHING AT ALL

9

> The Great Beginning produced emptiness and
> emptiness produced the universe.
> —Liu An, *Huai-nan Tzu* (second century B.C.E.)

In the modern, prevailing view of the cosmos, we sit here as tiny, unimportant specks of protoplasm, flukes of nature, and stare out into an almost limitless void. Vast nameless tracts of emptiness dominate the scene. Talk about feeling small.

But, as we will see in chapter 12, we do not look out at the universe; it is, instead, within us, as a rich 3-D visual experience whose location is the mind. Then what about all that supposed nothingness, that yawning lifeless gap between the stars and galaxies? Dispensing with empty space will take us most of the way toward discarding the notion of a little island "me" bravely forging ahead in a vast, lonely cosmos. It will help demolish the modern image of insentient vastness being the dominant quality of reality.

By contrast, the life-centered view is that "space" is largely a sense of order created exclusively by the mind's automatic algorithms. Beyond the observer, no real emptiness exists. For all these reasons, an exploration of space is important. It's also enjoyable for anyone who finds "nothing" fascinating. That's because, to start feeling comfortable jettisoning the existing mainstream view, the word *exist* proves to be a key, since our merry boat ride down the river named existence begins by examining its antithesis: nothingness.

The universe does seem a huge virtual ball of emptiness, according to physics texts. Yet even here on Earth, the richness around us is an illusion. Remove all the unoccupied space within each atom, and the entire planet would pack itself into the volume of a marble. This marble would then be a black hole, in that such density would sufficiently boost its gravitational field that its own light could not escape. It's a marble weighing six sextillion tons.

Beyond Earth, space is actually not as devoid of known material as the space within atoms is. Fanciful sci-fi writers sometimes suggest that a solar system with its orbiting planets is analogous to electrons whirling around an atomic nucleus. It's actually a bad metaphor. Relative to the sizes of their components, atoms are ten thousand times emptier than solar systems. Nonetheless, between planets and stars, very little is discernible to our eyes and telescopes. That does not, however, mean that it's empty; the truth turns out to be quite the opposite.

Figuring out the nature of space has obsessed humans ever since the earliest written records of *Homo bewilderus*. The ancient Greeks, compulsive logicians, argued that the blank-seeming sections of the universe couldn't be empty because nothingness cannot exist. They said that for space to "be nothing" requires us to take the verb *to be*—which means to exist—and then negate it. *Being nothing*, they said, is a contradiction. It makes as much sense as saying you're walking not walking.

During the Renaissance and its sudden proliferation of European and then American deep thinkers, most eighteenth- and nineteenth-century scientists said that light is composed of waves (Newton was a notable exception because he thought of light as particles), and waves require some medium through which to travel. Sound waves need air to go from a teenager's car radio to pedestrians hearing the thumping bass. Similarly, it was believed illumination waves from the Sun or the stars must require a medium to carry light's pulsations from there to here. The Church chanted "amen" to the "no such thing as nothing" credo—if God is omnipresent, there cannot be any vacuum. Thus the anti-nothing lobby included members of the scientific, religious, and philosophical communities. They ruled. You were in the nutjob category if you were pro-vacuum. The universal stuff assumed to fill all space was first called a *plenum*, then an *aether*, or *ether*. Its existence was a given for centuries.

The ether-belief only started to look iffy after one of the most famous demonstrations in history—the Michelson–Morley experiment, conducted in 1887. Albert Michelson argued that if the Earth was plowing through ether, then anyone on our world who aimed a beam of light in the same direction we are orbiting should see that light get a speed boost and reflect from a mirror faster than a similar light beam aimed at right angles to it.

To visualize why this would be so, imagine if the commissioner of Major League Baseball allowed a major exception to its rules and let a pitcher hurl his best fastball from the bed of a speeding pickup truck. The pitcher will throw the ball over the top of the cab toward the plate when the vehicle reaches the pitcher's mound, so the ball's release happens at the usual distance to the batter—60 feet, 6 inches.

If the pickup were to go 100 miles per hour at that point, and the pitcher released his fastball at 100 miles per hour, the batter would be staring at an incoming ball moving at 200 miles per hour. It would be very challenging to hit, to say the least.

In a similar way, nineteenth-century physicists assumed that if we beamed light in the forward direction of Earth's orbital motion, each photon would enjoy a 66,000-mph speed boost, compared with light we might aim sideways—or, especially, backward—from our track through space. Would there be a way to measure this effect?

With the help of Edward Morley, Michelson created an experiment using an apparatus with multiple mirrors sitting atop a stable concrete platform, floating on a pool of liquid mercury so it could be readily rotated. Knowing which way Earth travels, they first aimed light forward to where it would hit a mirror and bounce back; they measured the time interval it took. (The "bounce back" should be faster, too, just as a squash ball makes a quicker bounce back to you if you've hit it against the wall at a greater speed.)

Then the apparatus was rotated 90 degrees. Another pulse of light was flashed, this time to a mirror that did not sit in the "straight ahead" direction of Earth's motion. The results were incontrovertible. The light that traveled back and forth *across* the supposed "ether stream" that supposedly fills the entire cosmos, including every room in our homes, accomplished the journey in exactly the same time as light going the same distance forward in our planet's travel direction. Either Earth had stalled in its orbit around the Sun, or the ether didn't exist. (Yet another explanation—that light has a constant speed independent of everything else—was too weird yet to be entertained.)

Albert Einstein settled the matter a few years later. In 1905, his first relativity theory showed that light travels happily through a vacuum. Its waves are electric and magnetic pulses. Nothing was needed to convey them. This was welcome news. It hadn't really made sense for the planets to be passing through a substance without the slightest resistance. It was time to ax the ether with a good riddance.

Now fashion totally swung the other way, and "nothing" pleased everyone. Even the Church was no longer anti-vacuum.

Ah, but not so fast. Light from distant stars showed evidence that a small portion of it was being absorbed by some sort of wispy intervening material. Some skimpy stuff must be occupying space after all. Simple calculations revealed that, on average, one atom floats within each cubic centimeter of space.

Around Earth, the concentration of minuscule material is much higher, because the Sun sends out a constant stream of disembodied atom fragments. This "solar wind"—the term created for the phenomenon by physicist Eugene Parker in the 1950s, which was confirmed during the first satellite launches late that decade—has an average density of three to six atoms per sugar-cube-sized volume of space. It's substantive enough to push comet tails backward like airport wind socks and make them always point away from the Sun. Support for a small amount of absorbent floating or zooming material also came from the cosmic rays that continually hit our planet, discovered about a century ago. They presumably originate in distant, violent events like supernovae, as stars explode and fling their detritus wildly outward.

Despite all that, space is so uncrowded it wouldn't be wrong to call it a hard vacuum. So where is all this stuff we're saying fills every nook and cranny of reality? The key is that there's more to "space" than a mere recitation of its particle density. For starters, it's permeated by *fields*. Magnetic and electric fields fill the cosmos, and these have the power to influence the motion of every particle with an electrical charge. Space is also penetrated by an unrelenting torrent of photons of all kinds, which are the most prevalent entities in the cosmos. Neutrinos, the second-most common item, also continually rip through the entire universe; a trillion of them pass through each of your fingernails each second. And gravity waves flow everywhere, according to physicists. Thus, a lot is present even if it all weighs little or nothing.

Then there's the so-called dark energy that is making the visible cosmos expand. Its existence was unknown and unsuspected before 1998. That's when new measurements of cosmic distances, using a particular type of supernova to serve as a "standard candle" of luminosity, showed that the cosmos is getting ever more rapidly larger. Even in an explosion, the rapid outrush of material quickly slows down. But in the cosmos, the outrush is growing more and more energetic. This apparently began when the cosmos was half its present age, or about seven billion years ago. It's as if each group of galaxies has its own powerful rocket engine, and all of them suddenly turned on at that moment.

As this is clearly impossible, physicists groped for some explanation. Their best guess: dark energy. We know nothing about it, of course, except that it must be some kind of antigravity force. It would have to pervade all of space. It is theorized that, when the cosmic expansion had made the universe large enough, distances between galaxies were big enough to let this dark energy start to overwhelm the local gravity glue. The emptier the cosmos becomes, the more this dark energy can prevail, since the very emptiness of space is the home of this repulsive force. Moreover, because energy and mass are equivalent, and the amount of energy needed to blow apart the cosmos so enormous, this dark energy must be the predominant entity in the entire universe!

If this quality of space is the underlying cause of the Big Bang, then the universe is still banging—all thanks to its "empty" space. Thus, upon closer examination, it starts to look as if "nothing" is actually a vital, puissant *something*. Nowadays the cosmos is believed to be crammed with vacuum energy, seeming emptiness that actually seethes with unimaginable power.

Harder to grasp is an entirely different aspect of emptiness—one that has changed space from logical to enigmatic. We've seen that, especially since the late 1990s, experiments have confirmed the reality of entanglement, where two bits of light or actual physical objects, even clumps of material that were created together, fly off and live separate lives, but are always "aware" of the other's status. If one is measured or observed, its twin knows this is happening and instantaneously assumes the guise of a particle or bit of light with complementary properties. This "information" traverses empty space with no time lag, even if the twins are on opposite sides of the galaxy. In short, space is penetrated instantaneously, in zero time, no matter the distance.

All this strongly suggests that the gap between bodies is not real on some level. Emptiness is not what we once assumed it to be. If far-apart objects can be in simultaneous contact no matter the distance, what does this say about space or separation?

And if that weren't enough to establish a science-based connectivity between all objects, no matter their apparent separation, there's more. Einstein's special theory of relativity shows that space is not a constant and therefore not inherently substantive. High-speed travel makes intervening space dramatically shrink. Thus when we contemplate the cosmos, perhaps by gazing at the starry canopy during a camping trip, we may marvel at its distance and the universe's vast spaces. But experiments have repeatedly proven that this seeming separation between ourselves and anything else is subject to point of view—what Einstein called a *reference frame*—and therefore has no *inherent* bedrock reality. That's why Einstein himself did away with space as being any sort of trustworthy actual entity on its own, and replaced it with a mathematical concept of *space-time*. His revelation was that, taken on its own, space is tentative because it alters its dimensions. No gap between any two objects is reliable and inviolable.

Simply change your speed, or ask your real estate agent to find you a nice ranch house on a world with a much stronger gravity, and you'd find that all those stars now lie at entirely different distances. If we crossed a large living room going at 99.9999999 percent of lightspeed, every instrument and perception would show that it's actually now 22,360 times smaller—barely larger than the period at the end of this sentence. Space would have changed to nearly nothing. Where, then, is the supposedly trustworthy space matrix, the gridwork within which we observe the universe or even the objects in our earthly environment?

Beyond all this science, none of which is tentative or doubted by any physicist, looms the issue of whether gaps or separations exist objectively, or are merely the result of our minds' nonstop process of imparting order to what we see. Remember, we perceive only a limited range of electromagnetic wavelengths and only feel objects because our electrical fields are encountering theirs. Based on these sensations alone, we perceive seeming absences or empty gaps. Thus, apparent space is part of the mental logic of the animal organism, the software that molds sensations into multidimensional objects so that we can make sense of the world and accomplish all our vital functions, like finding food or searching for where we hid the TV remote.

When we think about it at all, most of us would probably regard space as some sort of a vast container that has no walls, which houses all visible entities. Myriad separate objects seem to lurk in this huge floorless warehouse. Seeing them as individual items requires that each object be identified as a pattern imprinted on the mind. Surrounding space is then required to identify them as separate entities.

But these gaps are often mere mental constructions. When we view a waterfall, do we count the spaces between droplets as gaps, or, instead, include it as the "waterfall object?" What about the mist—count it in, or out? How about the Sun? Would you include its interior as being "outer space?" Most of us would say no—the entire Sun is a single material body. And yet its gases and plasma are almost entirely empty spaces, especially within each of its atoms. So the more we ponder this, the more arbitrary becomes the notion of what is empty and what is not.

There's more. Actual nothingness would of course contain no oomph, no power. How could pure emptiness exhibit animation? And yet, for over a half century, astrophysicists have believed the universe's vast tracts of emptiness seethe with energy. As we'll see, this pathway will let us now get closer to grasping the limitless raw power of the mind-nature amalgam.

We've already seen that, no matter how cold and perfect a vacuum, it's still penetrated by starlight, infrared heat, and microwaves left over from the toasty Big Bang. These permeate the vacuum and need no medium to propagate. Because energy and mass are equivalent, those waves zipping through all of space mean you can't ever have a true vacuum.

But that's technical nitpicking compared to the real anti-nothingness news. German physicist Werner Heisenberg's uncertainty principle, published in 1927, claims a perfect vacuum can't exist. This was seconded at the time by other theorists who argued that empty space ought to contain a weird sort of energy. Back then, no one could find a trace of it, even if theory said that every cubic centimeter of blank space should contain more power than if each of the universe's atoms were the soon-to-be-created atom bombs. It took a while, but they were ultimately proved correct. Experimental evidence shows that "virtual particles"—things like electrons and antimatter positrons—snap, crackle, and pop out of nothingness everywhere all the time. Each particle typically exists for just a billionth of a trillionth of a second and then vanishes. If there's an energy field around, a virtual particle can borrow some of it to remain in existence forever. Thus, the seemingly empty universe forever swarms with exuberant, evanescent particles, like fleas jumping up and down on a hot griddle.

Physicists now believe this underlying "vacuum energy" is more than merely omnipresent; its power is enormous. Estimates of the energy in each small bit of seemingly empty space vary greatly. It's likely the space inside a mayonnaise jar contains enough power to boil away the Pacific Ocean instantly. (In truth, scientists are still in their early stages of understanding this all-pervasive energy. A disquieting 100 orders of magnitude exist between theoretical predictions of its power and measured values to date. This gap is known as the *vacuum catastrophe*.)

But although its true power is unknown, the existence of vacuum energy is scarcely in doubt. For proof, we first have the Casimir effect. It was named for Dutch physicist Hendrik Casimir, who made an odd prediction in 1948. He said if you hang two flat metal plates very close to each other, you'd limit the vacuum power between the plates because energy waves need elbow room. That's why ocean waves don't exist in sheltered coves, and why your eyes don't boil while you stare into the microwave oven—microwaves are too big to fit through those little holes in the door's screen. So the narrow gap between the plates restricts the wavelengths available for virtual particles. But the quantum energy outside the two plates is as strong as ever, and it pushes them together.

Well, this truly happens. The Casimir effect is real. Hang two plates apart by 100 times the width of an atom, and they refuse to just remain limp. Instead, they spontaneously move toward each other, pressed together with a force of 15 pounds per square inch. Move them twice as close together, and the force increases sixteen-fold. Something in empty space exerts a powerful force.

Some dreamers want to exploit the vacuum energy to give the world unlimited free power. Power from nothing. But there's a problem. This energy exists everywhere equally—which is why we don't sense it or detect it. Energy flows only from a place of greater energy to a place of lesser energy, just as heat moves only to where there's less of it. So how would you set up a condition that had less energy than that which is everywhere? How could you make it come to you, and therefore control it to create unlimited power?

The closest we get is when we chill matter to absolute zero at −459.67° Fahrenheit (−273.15° Celsius), where all molecular motion grinds to a halt. Then and only then are things at parity with this all-pervasive power. Hence, it's also known as *zero-point energy*.

There's evidence that this hidden geyser shows itself at that point. Helium couldn't still be liquid at absolute zero if it weren't receiving a bit of energy that keeps it from freezing solid. (It's the only element that doesn't naturally freeze, no matter how cold.) So zero-point energy makes itself present when all other energy is absent. To get this limitless quantum-foam energy to flow to you, you'd have to create below-absolute-zero conditions. This means making atoms move slower than "stopped."

Slower than stopped? The Greeks surely would be scratching their heads over that one. Solve it, and the power of the universe is yours. Meanwhile, let's be clear: That the cosmos is suffused with energy that makes the mere light waves and electrical fields around us seem by comparison like wimpy pretenders means that this essence of Being—this Nature of All Things, this true Self behind awareness and life itself, this seeming void that appears to be the matrix, the easel, the backdrop for all our human misadventures—is an unimaginably powerful entity.

Its energy is off the scale. Its potential is limitless. That we visually see and physically feel none of it means nothing; our senses are architecturally constructed to perceive what's useful in our everyday lives. What purpose would be served by perceiving the blinding ultra-energy that permeates every crevice of reality?

So, let's change our way of thinking of the cosmos. Let's regard visible objects as mere bits of flotsam materializing out of the vastly more powerful underlying vacuum energy, which is ignored because it's visually imperceptible. In this different mindset, might we perceive a fundamental oneness rather than individual entities separated by space? Can we not realize that, in any case, we block out known objects by our thinking mind within boundaries of color, shape, or utility? So is space always a reality, or is it a mere perception?

Let's sum all this up by first remembering the multiple reasons why space cannot be the simple blank gap between bodies assumed not too long ago. Shall we count the ways?

1. Empty space is never empty, especially when we include fields, photons, neutrinos, vacuum energy, and transient particle-pairs.

2. Distances between objects mutate depending on a multitude of relativistic conditions, so that no inviolable distance exists anywhere, between anything and anything else.
3. Quantum theory casts serious doubt about whether even far-apart bodies are truly and fully separated.
4. Separations between objects are often called space only because language and convention makes us draw boundaries.

Then, too, biocentrism shows us that since the observer and the universe are correlative, the space "out there" is part of a continuum of consciousness, and nothing exists apart from the observer. In reality, the farthest regions of space are located here in our minds.

Still, the mental torment imposed by the issue of space shows no sign of abating. Theoretical physicists wonder if there is a smallest possible amount of space that cannot be subdivided—some say yes. Others propose additional dimensions to space, beyond the three spatial dimensions and a fourth constituting time. There are complex, mathematically plausible arguments for extra, unseen dimensions. On the other hand, many scientists say that additional space dimensions are mere speculation, and must remain so unless some actual experimental or observational evidence comes to light.

Even crossing off the wacky-sounding stuff, we're left with a lot to think about. We started with a simple question—"What is space?"—the main component of the cosmos. And we end up with our heads spinning. One thing is clear: Our collective, long-held picture of the cosmos is disproved—and we can't even claim that this is a brand-new conclusion.

Even back in 1781, the same year the new planet Uranus was discovered, the Prussian philosopher Immanuel Kant wrote that "we must rid ourselves of the notion that space and time are actual qualities in things in themselves . . . all bodies, together with the space in which they are, must be considered nothing but mere representations in us, and exist nowhere but in our thoughts."

Biocentrism, of course, shows that space is a projection from inside our minds, where experience begins. It is a tool of life, the form of outer sense that allows an organism to coordinate sensory information and to make judgments regarding the quality and intensity of what is being perceived. Space is not a physical phenomenon and should not be studied in the same way as chemicals and moving particles.

Kant further said, "It is our mind that processes information about the world and gives it order . . . our mind supplies the conditions of space and time to experience objects."

In biological terms, the interpretation of sensory input in the brain depends on the neural pathway it takes. For instance, all information arriving on the optic nerve is interpreted as light, whereas the localization of a sensation to a particular part of the body depends on the particular route it takes to the central nervous system.

"Space," said Einstein dismissively, refusing (for the moment) to be embroiled in deeper philosophical thoughts about it, "is what we measure with a measuring rod." But it's clear that even his definition should emphasize the *we*. For what is space if not for the observer?

We could perform one of Einstein's classic thought experiments and try to imagine the cosmos if all objects and life were removed. Our first impulse might be to say, "Space alone would exist." But a moment's thought shows how empty (ha!) this demonstration is. For aren't we back to the ancient Greek railing against nothingness, how you cannot "have" nothingness? What would define its borders?

It is inconceivable to think of anything existing in the physical world without any substance or end. Even if actual emptiness still had a place in science—which, as we've shown, it no longer does—it would be meaningless to ascribe independent reality to truly empty space.

Thus, we are not individuals "here" with some empty gap—dead space—standing between us and, say, other galaxies. Space is unreal on multiple levels, and it is misleading to conceive of All There Is as some vast, mostly vacant sphere. The size of everything is dependent on reference frames, and further mutates via quantum laws so that it is questionable whether any absolute separation endures. A connectedness thus permeates what we used to call the ether.

Finally, in trying to answer the old questions about the size of the universe—now known to include consciousness and to be correlative with ourselves—we can only experience futility in any effort to "picture" an entity with no fixed dimensions.

So in addition to the cosmos existing outside of time, and having no death or birth, and seeing that *space* is a word that symbolizes nothing meaningful, we have arrived at yet another revelation:

The universe is sizeless.

THE RANDOM
UNIVERSE

10

"A Throw of the Dice Will Never Abolish Chance"
—Title of poem by Stéphane Mallarmé (1897)

We wouldn't struggle to invent the taco if it already existed. Why spin one's wheels on an unnecessary project? The only reason to create a new model of life and the cosmos is if the prevailing paradigm is faulty.

Is it? Well, we already saw in chapter 1 that the standard biography of the cosmos, recited in schools globally, involves a Big Bang followed by nature's four fundamental forces fashioning their magic upon two of the three varieties of fundamental matter. (That is to say, quarks and electrons; we can ignore neutrinos since they play no role in building the objects that make up the universe.)

Life and consciousness, according to this model, are central neither to the process of creation nor to its evolution or sustenance. They're afterthoughts. Accidents, if truth be told. That you and I are even here is a sort of trivial fluke. The arising of life is as inconsequential to the cosmos as the fact of Saturn's rings. It's a sort of pickle on the plate. An embellishment. As we life forms ponder the matter, we may regard life as a crowning feather in Nature's cap, but scientists concede it was hardly central or necessary in the cosmic timeline.

By now, the reader is well aware that *our* view couldn't be more antithetical, since biocentrism, as its very name suggests, means that life and awareness are indispensable cosmic attributes.

Proof that this is so is, of course, what this book is about. The case is not unlike those in a court of law, a step-by-step offering of evidence. And a vital step requires the disproving of the prevailing view. For so long as the current paradigm is accepted, alternatives will fare no better than to be consigned to the library's "what if" and "maybe" sections.

We have already seen that the existing view is firmly rooted in a space-and-time modality: You and I are bodies on a planet that dwells in a particular cosmic neighborhood. Our world had a birth 4.65 billion years ago, some 9.15 billion years after the Big Bang, and so on. This is how we visualize things, or, hopefully, how we used to, as we've already seen that neither space nor time has any kind of fundamental reality beyond being tools of animal perception. Once we've dispensed with space and time, one other major player has a central role in the current standard model: randomness, or chance.

We're all familiar with "the law of averages," and no one can dispute its value. We know that if you flip a coin ten times, the most likely result will be five heads and five tails. But we wouldn't be amazed if instead we got seven heads and three tails. Therefore in a single trial involving ten flips it would raise no eyebrows if heads showed up 70 percent of the time. If we took a statistics course in college, we'll also recall that a large sample size or N makes the law of averages truly start to appear so magical as to be almost carved in stone. Thus, if we flipped a coin 10,000 times, we could be very confident that heads would *not* appear 7,000 times, even though this result apparently duplicates the 70 percent heads outcome of that first experiment. Indeed, getting 7,000 heads would be so bizarre, we'd be wise to distrust the veracity of the coin or the impartiality of the experimenter rather than accept the result.

Statistics, in other words, provides a very trustworthy path when we wish to figure out what's happening. That's why, when subscribers to the "dumb random universe" model (meaning, almost everyone) state that absolutely everything arose by chance, it seems reasonable. Chance also makes it appear plausible that a cosmos as numb and insensate as shale could, given enough time, come up with hummingbirds by randomness alone.

Standing opposed to this are a spectrum of religious viewpoints; however, we will deliberately leave God out of this, especially since there are other conceivable non-chance ways the cosmos could fashion complex architectures—if, for example, Nature is innately smart, and intelligence is part and parcel of the whole shebang.

Some version of inherent cosmic intelligence or else Creator-Deity was assumed for countless centuries; it was the prevailing, almost invariable mindset of scientists, who were called *natural philosophers* until the nineteenth century. Even as brilliant a thinker as Isaac Newton wrote, near the end of his life, "Whence arises all that Order and Beauty which we see in the world? . . . How came the bodies of animals to be contrived with so much art? . . . Was the eye contrived without skill in optics?"

Or one can turn to Cicero, who wrote, over two thousand years ago, "Why do you insist the universe is not a conscious intelligence, when it gives birth to conscious intelligences?"

So the "smart universe" paradigm prevailed through most of recorded history, either by acknowledging an omniscient puppeteer—God—or by assuming the acumen to be innate, as in, "You can't fool Mother Nature." Completely eliminating cosmic intelligence in any form is a rather recent development, even if it is the current science norm. Still, in popular parlance, folks continue to say things like, "Nature knows what it's doing."

In any event, the modern dumb-universe paradigm requires that we explain the complex physical and biological architecture we see all around us by some other means. And chance is all we have. It's all an accident. The dumb-universe model sinks or swims on the life raft of randomness.

Randomness is also a central key of evolution, where it works splendidly. Darwin wasn't whistling in the wind with his natural selection. It makes sense that giraffes developed long necks because those giraffean predecessors who by chance had received a random mutation for a longer than normal neck had a survival edge when it came to grabbing leaves and fruit from higher branches. Over time—and it doesn't take terribly long—the preferential breeding selection of longer-necked mammals gave them a leg up in the Serengeti.

Evolution works, and it's based on random mutations coupled with natural selection. This being so, the science community is happy that the public lazily considers "chance" applicable to everything else we see, too. This includes the entire universe and the rise of life and consciousness.

Now, many if not most fundamentalists and intelligent-design, Bible-based groups deserve their reputation for being obstinately antiscience. They defend the Bible at all costs, even when it claims that a person named Noah saved two members of every species, of which there are eight million, to survive a worldwide flood for which no evidence exists. (Indeed, aside from the fact that two animals wouldn't provide enough biodiversity for a species to survive, a global flood deep enough to submerge the Himalayas is problematical, since only a one-inch sea-level rise would ensue if every molecule of Earth's vapor precipitated out as rain.) Their defense of Scripture, no matter how far-fetched the particular passage, handcuffs them to untenable positions. But give them this: When they complain that the creation of the eye's architecture cannot be explained by natural selection, and some scientists respond by summarily dismissing them, it is the latter who are guilty of sloppy reasoning.

Natural selection works because some random mutation conferred an advantage that let the animal better survive to procreate. But an eye—any eye, even the earliest ones—required not just a single mutation that created a light-sensitive cell, but also a nerve system or some other modality to carry such sensations to a brain or brain precursor, so the information could be utilized in some way, such as locomotion toward or away from the light source. Sight also requires a "perceiving" cell structure in which to form an image, even if it's just a crude sensation of brightness. In short, even primitive vision involves far more than a single genetic mutation. No matter if the earliest eyes lacked the sophisticated elements of current animal vision, with their marvelous supporting cast of muscles for focus and adjustable pupil diameter; various types of color-sensing retina cells; lens; optic nerve; and an amphitheater of billions of specialized neurons and synapses to actually create image perception. It's quite an elaborate architecture that today's animals enjoy. But even the first, crudest version would require *some* structure to be the least bit useful.

A single mutation would accomplish nothing. It would confer no benefit, and thus there'd be nothing advantageous to pass on to the kids. And what are the chances for a profusion of simultaneous, independent, but interdependently necessary mutations occurring in a single animal?

Thus (goes one argument) the eye, and several other complex biological arrangements in which the components do not work unless an entire architectural structure is in place, are all evidence for an innate "design" intelligence or else (as they believe) a skilled Creator. In short, evolution beautifully explains the improvements in species along with adaptive strategies and configuration changes, but it doesn't explain many of the *original* biological facets like the initial arising of life, or even some vital organs.

There's another problem with lazily letting evolution be the explanation for virtually everything that concerns life and its changes. Although classical evolution does an excellent job of helping us understand the past, it fails to capture the driving force. Evolution needs to add the observer to the equation. Indeed, Niels Bohr, the great Nobel-winning physicist, said, "When we measure something we are forcing an undetermined, undefined world to assume an experimental value. We are not 'measuring' the world, we are creating it."

The evolutionists are trying to pull themselves up by their bootstraps. They think we, the observer, are a mindless accident, debris left over from an explosion that appeared out of nowhere one day. Loren Eiseley, the great naturalist, once said that scientists "have not always been able to see that an old theory, given a hairsbreadth twist, might open an entirely new vista to the human reason." The theory of evolution turns out to be the perfect case in hand. Amazingly, it all makes sense if you assume that the Big Bang is *the end* of the chain of physical causality, not the beginning.

If we, the observer, collapse these possibilities (that is, the past and future), then where does that leave evolutionary theory as described in our schoolbooks? Until the present is determined, how can there be a past? The past begins with the observer, us, not the other way around as we've been taught.

Although the preceding might take a while to sink in, what's immediately inarguable is the futility of assigning randomness as any kind of genesis in the development of *consciousness*. The fact of having perception, of being aware, is a quality that has eluded all researchers. Its birth defies even the simplest guesses. Indeed, those who have studied it join Ralph Waldo Emerson in declaring it a profound mystery akin to peering at "a holy place." This enigmatic quality sets up barriers and challenges for the scientist, since everything we see and think about the universe—the very act of seeing and thinking—involves perception. If awareness contains its own built-in biases—and we will show that it does—then we cannot begin to understand the cosmos without first grasping consciousness itself.

But let's not get too far afield. Standing apart from all this, from consciousness and biological life, is the modern existing paradigm of universe construction whose cornerstones are time, space, and randomness. We've carefully explored and demolished time and space as independent self-existing entities. Let's be equally thorough when it comes to chance.

As observers, we assume that random events created most or all of what we see. The cratering pattern on the planet Mercury appears as random as a jackal's markings. And in the quantum world of the tiny, we only understand things probabilistically. Whereas in many areas this works splendidly, "chance" is actually a fascinating process that's often misunderstood.

The most famous illustration of probability is the monkeys-and-typewriters thought experiment. We've all heard it. Let a million monkeys type randomly on a million keyboards for a million years, and you'd get all the great works of literature. Would this be true?

About ten years ago some wildlife caretakers actually put out a computer and keyboard in front of a group of macaques to see what would happen. The animals typed virtually nothing. Instead they threw the keyboard on the ground, used it as a toilet, and quickly rendered the apparatus useless. They didn't create any written wisdom whatsoever.

But let's get serious. We'll confine the experiment to our minds the way Einstein liked to do in his thought experiments. So *could* a million diligent monkeys typing for a million years truly create *Hamlet*? And if one of them wrote *Moby-Dick* word for word on her ninety-seven-billionth attempt at pounding random keystrokes but then left out the period at the end, would that count?

Believe it or not, such a problem is entirely solvable. Now, keyboards offer a lot of places to push; let's say each type-writer has fifty-eight keys. When talking about random events, consider the difficulty of creating merely the fifteen opening letters and spaces of *Moby-Dick*, "Call me Ishmael." How many random tries would be needed?

Given fifty-eight possible keys, it would be 58 × 58 × 58 × 58 . . . fifteen times over, which is about 283 trillion trillion attempts. But remember we have a million monkeys working, and let's say they type forty-five words a minute, so the fifteen keystrokes that make up the phrase take just four seconds. And they never rest or sleep. How much time, then, according to probability laws, before one of them finally types, "Call me Ishmael"?

Answer: about 36 trillion years, or roughly 2,600 times the age of the universe.

So a million monkeys typing furiously would never even reproduce one book's single, short opening line. Moral: Forget the monkeys-and-typewriters thing. It's bogus.

The real problem with reliance on chance to explain what is otherwise unexplainable is that it far overstates the power of random events. For example, astronomers certainly hope to find life elsewhere and would automatically assume that any alien life form's existence would have initially arisen through random physical or chemical processes. Using this assumption, exobiologists might then attempt to solve the issue of life's genesis in that remote star system. But our point is that the random supposition is simply not any kind of useful hypothesis. Since the random business is given far more potency than it deserves, both in the popular imagination and among scientists, we'd be more likely to make progress by candidly saying, "This is a mystery"—and then researchers might begin to tackle it from scratch with a clean slate.

Accomplishing some particular complex task by mere chance—like the creation of life and consciousness—is what we're examining here. Given the stupendous limitations in what chance can accomplish, we must also understand why—seemingly paradoxically—random events do nonetheless create a dizzying array of possibilities.

Consider the ways you can arrange four books on a shelf. You find the possibilities by multiplying 4 × 3 × 2—which is pronounced "4 factorial" and written 4!—which amounts to 24. But what if you have ten books? Easy again; it's 10 factorial or 10 × 9 × 8 × 7 × 6 × 5 × 4 × 3 × 2, which is—ready?—3,628,800 different ways. Imagine: Going from four items to ten increases the possible arrangements from 24 to over 3.6 million.

Let's picture this. We can easily imagine taking ten books out of some storage box and then quickly putting them on a shelf haphazardly. Would we ever guess that the chances are about 3.6 million to one against them appearing alphabetically? Few of us would imagine such long odds. Sure, it's very unlikely that they'd just happen to land alphabetically. But 100 to 1 sounds more plausible. A thousand to 1, tops. Over three million to 1 doesn't seem realistic. Yet it's true. That's the same as putting up those ten books every single day for over 100 lifetimes, before you'd achieve that arrangement.

Possibilities are always insanely enormous. They surprise us. The number of atoms in the entire visible universe can be written right here: 10000000000000000000000000000000 00000000000000000000000000000000000000000000000 00—that's eighty zeroes. Add just six more zeroes (you'd hardly notice them) and you've represented *all the atoms in a million universes.*

But you'd have to type zeroes for the rest of your life to express the ways—just representing them in writing—that stars can be arranged in our galaxy. Or that neurons can connect in a human brain. The number of ways things can happen is stupendous. The mind's potential lies beyond its own comprehension. (One of our favorite quotes is from George E. Pugh: "If the human brain were so simple that we could understand it, we would be so simple that we couldn't.")

We can always count *things.* No problem there. But when it comes to assessing *possibilities*—on Earth or off it—we monkeys haven't got a chance.

So back to our original question: Can you get the cosmos we see, including the complex biological designs of the brain and the trumpeter swan, through random atom collisions alone? If randomness requires thirty-six trillion years to type a single passage of fifteen letters and spaces, the answer is obvious: not a chance. On the other hand, if the desired endpoint is not some specific accomplishment like mangoes or the genesis of life, and you're merely asking those colliding billiard balls to come up with something or other, anything at all, it will surely oblige.

This takes us inescapably to considering chance as it tries to create *some sort* of universe. The problem is, *our* universe has an exquisite set of properties that are Goldilocks-perfect for life to exist. We live in an extraordinarily fine-tuned cosmos. It's a place where any random tweaking that conjured even slightly different parameters in hundreds of independent ways would not do the job of allowing any kind of life to arise. Let the gravitational constant be 2 percent different, or change the power of the Planck length or Boltzmann's constant or the atomic mass unit, and you'd never have stars, or life.

So by any stretch of wishful thinking, a cosmos that even allows life—let alone the fact of life's development—is inconceivable by chance alone. *Randomness is not a tenable hypothesis.* Truth be told, as an explanation it's close to idiotic—right up there with "the dog ate my homework." It's almost as if "dumb cosmos" supporters demand their theory's validity to march in sync with its central premise.

And so falls the final cornerstone of the current "clarification" of the cosmos. Chance goes down the drain to join its comrades time and space. The modern popular model, which revolved around that triad, always seemed a sickly, forced explanation that requires little more than a cursory inspection to be demolished.

Of course, even had friendly background conditions and favorable physical constants all come into existence, life and consciousness—according to the modern paradigm—must still duly manage to arise purely by accident. These are not trivial, easily manufactured items.

Let's sum up the most basic do-or-die physical background conditions for life to spring into existence. First, two specific fundamental forces—electromagnetism and the "strong force," which operates only in very small spaces—must have specific values. The former permits electrical fields that can keep electrons attached to atomic nuclei, allowing the existence of atoms. But even atomic nuclei won't hold together without a perfectly tuned strong force, since this alone lets multiple protons cling together and overcome the like-repels-like nature of electromagnetism. Without multiple protons, the only element that could exist would be hydrogen. And although no one is against hydrogen, it alone could not produce any sort of organism, even if nature patiently waited eons until the cows came home.

Then you need a third fundamental, the gravitational force, to be not too weak and not too strong or you can't have stars. We could keep going, but suffice it to say that several dozen (some say as many as two hundred) physical parameters must be exactly as they are to within a percentage or two for stars to undergo nuclear fusion and create all their nice warmth and sustenance, for planets to form, and for multiple elements to be created. In short, yes, it's a perfect universe—and we haven't even yet gotten to the life-creation process with its own crowded stadium of requirements, such as worlds that are not too hot or cold or radiation filled, and specific properties of a few key elements like oxygen and carbon that need to exhibit just the characteristics we observe.

Even locally, here on Earth, life would be difficult or impossible if we didn't possess our massive nearby Moon. That's because our world's axial tilt would naturally wobble wildly, sometimes aiming straight at the Sun so that it would be overhead for months at a time, producing impossibly hot temperatures. But our planet manages to avoid going through such chaos. Our axis's obliquity is essentially stable and displays small harmless variations of ±1.2° around an average of 23.3°—just about where it's aiming today. If the Moon's gravitational torque were absent, the axis would change from nearly zero (meaning, no seasons at all) up to about 85°—meaning, aimed sunward the way poor Uranus does.

Thus the Moon regulates our climate, keeping it gentle and relatively consistent over the eons, instead of us periodically having impossibly hostile conditions that would have made ice ages seem by comparison like subtle room-temperature changes.

And how did we get the Moon? The perfectly timed collision of a Mars-sized body coming from a propitious direction and at the correct speed—not too fast or massive to destroy us, and not too small to fail to do the job. Direction matters because unlike all the other major moons of the solar system, *ours is the only one that doesn't orbit around its planet's equator.* Our Moon ignores our axial tilt. If it orbited normally, it wouldn't always sit in our orbital plane and thus exert its torque in a Sun-vector alignment, where it's maximally effective at stabilizing our axis. Another accident.

This is an extremely unlikely universe. So unlikely that even the most die-hard classical, randomness-believing, atheism-proselytizing physicists concede that the cosmos is insanely improbable in terms of life-friendliness. The combined existence of all the life-friendly values of all its physical constants and values defy the odds of one in several hundred million.

The following figures illustrate a few of the ways our reality is extremely improbable. Taken alone, each might be brushed aside. But considered as an aggregate, these "coincidences" produce a universe so astonishingly life-friendly, the situation demands an explanation.

Was our universe created randomly, by chance? If so, we repeatedly defied the odds. Ours is an *extremely* unlikely reality. The Sun—central to life—would not exist if any of several of the universe's basic physical constants were even a paltry 1 percent different from their actual values.

Had our Sun been significantly more massive, it would have blown up into a supernova long ago. Even having a massive star in our celestial neighborhood would have changed Earth's radiation flux when it "went supernova."

Earth has been hit by celestial objects, but none large enough to destroy it. It would have been a very different story if massive Jupiter didn't exist, gravitationally deflecting or altering the orbits of most incoming hazards.

There would be no stars and no life anywhere, and no element other than hydrogen, if the strong force inside every atom were just slightly weaker than it is.

Our lush earthly life would be impossible without the Moon. Its influence stabilizes our degree of tilt, preventing chaotic changes that would have made our planet inhospitable.

Our luck didn't stop with the physical properties of the universe. S. tchadensis, O. tugenensis, A. ramidus, A. anamensis, A. afarensis, K. platyops, A. africanus, A. garhi, A. sediba, A. aethiopicus, A. robustus, P. boisei, H. habilis, H. erectus, and H. georgicus—among other hominid species—all went extinct. Even the Neanderthals went extinct. We alone made it.

This hyper-unlikely nature, just on a strictly physical level, makes many physicists sigh with discomfort and admit that some sort of scientific explanation is badly needed. In turn this has provided a major motivation for the pursuit of ideas such as superstrings, to which some stubbornly cling, even though the current consensus is that it's a failed theory. String theory did more than provide a hope for fashioning a unification for all the forces and such. A mere two decades ago, there was optimism that by mathematically incorporating eight extra dimensions, it might explain why the cosmos is the way it is.

It doesn't. And it hasn't. To the contrary, string theory allows at least 10^{500} "solutions," so that its detractors dismissively call it a *theory of anything*. (And any hypothesis that allows anything actually explains nothing.) The reason it remains attractive to those who desperately want to explain the improbably life-friendly nature of our universe is that some of its few remaining adherents say that all those solutions are not evidence of a useless anything-goes hypothesis, but instead support the idea of countless *multiverses*—other parallel universes where all the endless string solutions manifest themselves.

How can this possibly help? Well, goes this reasoning, if there really are 10^{500} other universes out there, each with different random properties, then the vast majority will have physical laws that are not life-friendly. A few of these multiverses would, by chance, happen to harbor conditions that permit the existence of life. We live in one of those. Where else could we live, if we're here asking questions? Thus our own cosmos, with its seemingly impossibly life-friendly conditions, becomes not so strange. It no longer demands any sort of explanation. This string-based multiverse reasoning instantly lets our hyper-unlikely friendly universe experience a sudden metamorphosis and go from extraordinary to worthy of no more than a shrug. By such reasoning, the random explanation for reality gets a new lease. And lifelessness becomes the cosmic normal.

Naturally, most physicists aren't buying it. Columbia University mathematical physicist Peter Woit pulls no punches. "Physicists had huge success in coming up with powerful compelling fundamental theories during the 20th century," he explains,

> but the last forty years or so have been difficult, with little progress. Unfortunately, some prominent theorists have now basically given up and decided to take an easy way out . . . They allow theoretical ideas like string theory that have turned out to be empty and consistent with *anything* to be kept alive instead of abandoned. It's a depressing possibility that this is where physics ends up. But I still hope this is a fad that will soon die out. Finding a better, deeper understanding of the laws of physics is incredibly challenging, but it's within our capability as humans, as long as the effort is not overwhelmed by those selling a non-answer to the problem.

Applying Occam's razor—the theory that the simplest explanation is usually the best—we find that biocentrism offers an obvious alternative explanation for our undeniably improbable life-friendly universe. Namely, that it's life-friendly because it's a life-created reality!

With all this, let's *still* not assume reality contains any kind of underlying intelligence as opposed to mere dumb randomness. Instead, let's take a clean sheet of paper and continue to review what science has been actually telling us during the past century without any bias one way or another. Doing so, we'll now take a little side road.

FACING REALITY

11

You throw the sand against the wind,
And the wind blows it back again.
—William Blake, "Mock on, Mock on, Voltaire, Rousseau,"
The Notebook of William Blake (1796)

This book has already demonstrated that the universe is not the way it's commonly perceived. Science, logic, and the discoveries of the past fifty years show that our shared assumptions about reality are far from the truth.

But now let's take a side trip. In this chapter we'll see why this quest's conclusions also endure at a gut level, outside logic and science . . . how they're part of a grand tradition dating back countless centuries.

After all, if we are to be honest, clever enough phraseology can appear to prove anything—just as Zeno of Elea "proved" that, in a race, you could never overtake a tortoise. The authors are under no illusion that some readers will end up shrugging off all rational arguments and evidence. Thus, let's take a few minutes to briefly go in a very different direction. We'll explore a more intuitive approach, even if it does bypass the safe harbor of logical analysis.

It will surprise no one that our detour involves a turn to the East. It is there, in Hinduism and Buddhism, that these very issues remain front and center. This actually constitutes a major difference between Western religions and those with roots in the Indian subcontinent. In the Judeo-Christian tradition, duality is central to the perception of reality. The basics of life and the cosmos involve relationships—often encompassing tension or conflict—between the individual versus nature or the individual self and its relationship to a deity that is separate. They're almost always temporally structured, as when one's present life stands opposed to its spiritual goal, which supposedly lies in the future. Thus, for Westerners, a bedrock fundamental is the existence of time. Throw in the central mandated tenets of obedience, correctly practiced ritual, and rules for divinely sanctioned moral behavior in everyday life, and you've got the ingredients for most of the chapters in the Talmud, Bible, and Koran.

In all these, the universe had a beginning. God alone stands apart from time. Thus, His creation—Everything—exists in a time-based matrix. Time plays a central role in how we "should" be living and what we should hold most sacred. That's because all the good stuff, including our rewards for proper behavior, will only happen in an afterlife. And an afterlife is not now. It's later. Thus our traditions revolve around seeing everything in a time-based configuration. Doing so, we divide the cosmos into various spatio-temporal parts, of which our soul and body are just one minor piece.

This mindset seeps into all areas of life. We stare at an aurora or look through a telescope, and the most commonly heard comment is, "It made me feel so small." And although such humility seems admirable on paper, a much more uplifting perception would be feeling oneself to be absent altogether. Neither small nor large, but simply gone. Then alone, without the diversion of trying to be simultaneously aware of the observer, can the full experience of the perceived object manifest itself without distraction.

Contrast our dualistic worldview with that of the East. One can grasp the latter by perusing books—some written long before the Bible—or through the works of modern interpreters such as Paramahansa Yogananda, Ramana Maharshi, or Deepak Chopra, but it essentially comes down to this:

Whatever you think, however your logic works, Eastern sages have always insisted that there exists a direct experience of reality that is nonverbal. Eastern religions are thus *experience based*. Or, if you like, *hands on*. This lies in stark contrast to Scriptural wisdom, which is always secondhand even if the source is trustworthy. Scriptural wisdom is fine, but there's no substitute for seeing something for yourself. A book can warn you that a wood stove is hot, but just a single accidental touch and you'll never need to read anything more about it.

Some thirteen hundred years ago, in India, the since-revered Shankara wrote, "I am reality without beginning . . . I have no part in the illusion of 'I' and 'you,' 'this' and 'that.' I am . . . one without a second, bliss without end, the unchanging, eternal truth. I dwell within all beings as . . . the pure consciousness, the ground of all phenomena, internal and external. I am both the enjoyer and that which is enjoyed. In the days of my ignorance, I used to think of these as being separate from myself. Now I know that I am all."

When it comes to a direct experience of the very nature of reality, the event essentially boils down to seeing unity and peering through the illusions of time and death. It is variously called realization, enlightenment, union with God, satori, samadhi, nirvana, and many other names. Supposedly it's not just saints or some gurus who have had this transformational experience through the ages and throughout the world. Ordinary people have as well.

The reason we are even "going there" in this chapter—and committing a kind of science no-no by temporarily leaving empirical evidence for an anecdotal account—is because one of the authors (Berman) actually had this experience when he was twenty. This has produced a rather unique and interesting situation in the book's coauthorship. One author comes to biocentric conclusions strictly through science and logic; the other, despite fully agreeing with the science, primarily subscribes to the view on a gut level. Thus it seemed that instead of being coy about the topic of "direct experience of reality" and strictly quoting others who have written about it, we ought to share a firsthand experience, as recounted by Berman in 2008:

We trust our instincts. We need no textbook to teach us to love, or to recognize danger, or to be swept into joy by a beautiful garden. Yet when it comes to grasping the nature of existence, we fumble and stumble through insensate theories, our eyes glazed over as we hear about string theory's extra dimensions.

Life customarily offers disparate sources for knowledge. But how about the big-ticket issues of cosmology and existence? What's the correct tool there? Logic? Math? Science? Religious texts? Instinct?

I found out soon after I turned twenty. I'll share it now for the first time.

I was in my junior year of college, cramming for a test. I had breezed through most of my astronomy courses but, philosophically, the universe was still essentially a vast, mysterious entity. I had tried meditating during the past month, but couldn't really say I'd experienced anything revelatory. Now I was studying for a physiology test when something in the textbook about the visual part of the brain suddenly gave me a split-second insight that the distinction between "external" and "internal" is unreal. Then that intellectual insight abruptly changed into something else.

An enormous weight I'd never realized I had borne was suddenly lifted. An experience began that no words could convey. It was ineffable and life-altering. The best I can say is that "I" was suddenly gone, replaced by the certainty of being the entire cosmos. There was absolute peace. I knew with total confidence, not logically—because, as I said, Bob was no longer present—that birth and death do not exist. That all is perfect eternally, that time is unreal, and that all is one. The joy was beyond anything I could have imagined. The to-the-marrow certainty could perhaps be better described as a *recognition*, an ancient familiarity of being Home.

When the intense initial experience faded, the room returned, and my textbooks lay before me. Except, all was now profoundly altered. Let's call this "the second level of the experience." There was still no sense of a separate "me," an observer looking out upon the world. Rather, everything was a oneness, and I was whatever my eyes gazed upon. It was as if my consciousness had previously been long confined, like a canary in a little cage, and that a false sense of being a separate, isolated, thinking individual had now vanished. Objects were no longer separate items existing in space; instead, everything was the same continuum.

When a person came into view, I *was* this person. The universe was one entity for all time. There were not billions of humans and animals. There was *one* living, deathless entity. (And no, in case you're wondering, this experience was *not* chemically induced.) If this sounds fabulous, well, no words could begin to convey the clarity.

This experience lasted three weeks, during which time no thought flitted across my consciousness. But eventually the ongoing stream of mental chatter, of being an individual, an observer, returned—accompanied by loss of the peace and oneness. It felt terrible.

Afterward, I went overseas, mostly to the East, traveled in thirty-five countries. I tried everything, read spiritual books.

There were times of recapturing that lower "second level" of perception, but never again that full experience. Those spiritual books said that people in all cultures through the ages have had the same experience, and that it has variously been called enlightenment, awakening, and so on.

Indeed, nearly everyone has had moments, perhaps when watching something in nature, when one feels a rush of ineffable joy, of being taken "out of oneself" and essentially becoming the object observed. On January 26, 1976, the *New York Times* magazine published an entire article on this phenomenon, along with a survey showing that at least 25 percent of the population has had at least one experience that they described as "a sense of the unity of everything." It's apparently not that rare.

That's our coauthor's personal account. If it's delusional, then it's odd indeed that it mirrors accounts from different centuries and cultures. Such accounts also bring up a very different issue: What could possibly conjure up such a change in perception? How can neural circuits alter so profoundly as to create an entirely different universe, one at odds with everyday paradigms?

We already know that certain psychedelic drugs seem to do the job, albeit unreliably, since most people who take them have no such experience. Head injury, congenital brain anomalies, as well as techniques like yoga practices seem capable of altering the state of perception, too.

One of the authors (Lanza) explains it this way:

"All you have to do is change the data input and its interpretation by the detector (the brain and its complex neural perceptive system) and you perceive reality differently. Thus, we cannot trust our primitive animal brains to paint an accurate picture of what's *really* going on. Regarding the 'single entity' experience—this interconnectedness is consistent with the *global quantum state* (which we'll explore in chapter 19). If one could experience all knowledge—everything possible (i.e., everything that can be experienced in space and time)—our individual separateness would melt away, which is what happens in the entanglement experiments, and what these mystics seem to be describing."

The takeaway here is that you can restructure the neurocircuitry of the brain so we experience oneness rather than separateness. That it does happen spontaneously, but not to most people, may simply mean that such widespread perceptions would not be evolutionarily adaptive. Having everyone sitting around smiling and at peace might not be consistent with the nature of life, as it would produce very different choices and evolutionary pathways from the current modality.

For those who have not had such an experience or are skeptical about it, we can all probably agree that the mere fact that some presumably credible people report it is evidence for one inarguable thing: that our neurocircuitry can be very easily tampered with. In turn this illustrates just how subjective our view of the world really is. The cosmos itself mutates with biologic tweaking.

Lanza recalls, "In medical school I remember a patient who'd had a terrible accident, where a metal rod went into the visual portion of his brain. He went blind and couldn't see anything. Yet, if you put a horizontal pole in front of him he would duck even though he couldn't see it."

Such incidences are now called examples of "blindsight," and serve as another example of how deeply entwined are the neural circuitries that comprise our realities, and how they create the universe—essentially define reality—in ways we are only beginning to grasp.

On December 22, 2008, the *New York Times* gave front-page coverage to this phenomenon, by reporting on a man whose two successive strokes left him totally blind. Here the issue became: Is perceiving the visual world the only way we can see?

A neuroscientist at Harvard tried something remarkable. The patient was asked to attempt an obstacle course. Reluctantly, he agreed, and what unfolded was astounding. "He zigzagged down the hall, sidestepping a garbage can, a tripod, a stack of paper and several boxes as if he could see everything clearly," explained the researcher, who followed closely behind him in case he stumbled.

What we "see" is a complex construction generated in our head. One of the best proofs of this is the neurological phenomenon called "blindsight." These patients are blind due to injuries or lesions in the striate cortex of the brain. Although blind, they can navigate an obstacle course and even recognize fearful faces.

"You just had to see it to believe it," said the Harvard neuroscientist, whose paper appeared in the journal *Current Biology*, along with extensive brain imaging. In other words, we have an innate ability to sense things using the brain's primitive subcortical system, which is entirely subconscious. It's a visual system but it bypasses the usual visual pathways of the brain, and it employs other modalities than the normal images involving light and color.

This newest study, the first to show the blindsight phenomenon in a person whose visual lobes were completely destroyed, forces a conclusion that should already be obvious. The cosmos is perceived, and becomes what it is, based on our neural circuitry.

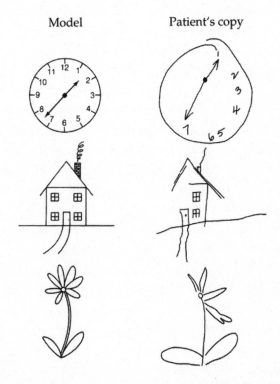

Model   Patient's copy

Numerous medical disorders demonstrate how subjective is our view of the world. Any tampering with the brain's hardwiring/neurocircuitry can radically change our sense of reality. For instance, in this case, a patient with hemispatial neglect (resulting from damage to the right parietal lobe of the brain) only perceives one side of the world and ignores the other half when doing tasks. The drawings on the right were made as the patient attempted to reproduce the models on the left.

Blindsight may be one more example of something we could call *implicit knowledge.* Implicit knowledge denotes helpful information that exists below the fully conscious level, yet is used routinely for such everyday tasks as walking and moving without slamming into things, making snap decisions, and communication with others both verbally and in texts and such. It is not necessary for one to have brain damage that renders them cortically blind to experience it; even normally functioning brains exhibit affective blindsight.

Put another way, people respond to stimuli and even subtler emotional information without having any conscious awareness of the process. A person may "reflexively" duck in a playing field if a football is about to hit them, exhibiting a level of perceiving that's extant beyond the standard channels of vision.

When cortically blind adults are shown pictures of scary faces or happy ones, it is accompanied by a measurable activation in the amygdala, the part of the brain associated with emotional processing. What's interesting is that everyone else—those with no brain damage at all—exhibits similar amygdala reactions when such evocative emotional images are presented to them at speeds far below the threshold of conscious awareness.

The bottom line is that blindsight—perceiving outside the normal physiological pathways—is available to everyone. Even animals. In 2015, researchers found that at least one octopus species can sense light without the help of the eyes or the brain.

Fine, the reader may be thinking, perception depends upon various brain mechanisms, even those that we are only just learning about. Still, isn't there a visual universe "out there" that exists independent of our biocircuitry? Aren't the sunset colors and blue sky self-existing, awaiting the clear-glass windows of one's eye-lenses, and the occipital-lobe visual receptors within some conscious animal, in order to perceive and enjoy them? In what way do these aforementioned experiences prove the unity of the subject and the natural world?

Of the many aspects of biocentrism, this is fortunately the easiest to demonstrate. For of all common misconceptions, the assumption that we look out upon the world is the most readily disproved.

WHERE IS THE UNIVERSE LOCATED? 12

"Here, There and Everywhere"
—John Lennon and Paul McCartney, song title (1966)

For some animals, the sense of touch or smell is paramount. For others, hearing is critical. Just watch Rover's ears as they swivel around. But humans rely on vision. In our explorations of the celestial realm beyond our planet, we have nothing else. We cannot hold the universe, nor can we smell it. Space is utterly quiet, so that the collision between small asteroids and the tumultuous births of galaxies unfold in silence. For us, knowledge of the cosmos arrives solely on the wings of photons.

We have known for a century that light is composed of waves of magnetism along with electrical undulations traveling at right angles to it. Neither magnetism nor electricity have inherent color or brightness, and thus even if there were an independent universe beyond consciousness, it would have to be utterly invisible. This bears repeating: At best, any separate external universe must be blank or black.

Yet look around. We're imbedded in a world of profound color and beauty. People assumed, until the advent of quantum mechanics a century ago, that our eyes' lenses were like clear glass windows that let us accurately perceive what is "out there"—and this remains the general public view even today. However, since we know beyond any doubt that what's "out there" can be no more than invisible magnetic and electrical fields, it's obvious that we ourselves—our neural circuitry—create the colors and patterns.

The biological mechanisms responsible for vision were researched for centuries, with many wrong turns alternating with "Eureka!"-like triumphs. Early philosophers rejected any notions that color and light were involved with an external world. Rather, wrote Plato in the fourth century B.C.E., light originates from *within* the eye, "seizing objects" with its own rays. But six hundred years later, the famed physician Galen disagreed, saying that vision is a function of an optical *pneuma*, meaning it flows from the brain to the eyes through hollow optic nerves. This idea of the brain being central to sight put Galen's perception fifteen centuries ahead of anyone else's.

Today, every physiology text paints a clear explanation for what we see "in front of us." First, light enters the quarter-inch-wide lens of each eye, where an upside-down image is focused upon the two retinas. There—at least in bright light, since dim-light vision employs different machinery—six million cone-shaped cells, which come in three varieties, each sensitive primarily to light's primary colors of blue, red, or green—are stimulated only when they receive the impact of a specific range of energy wavelengths. Upon stimulation, they send electrical signals up heavy-duty cables to an astounding universe of neurons designed to create three-dimensional images.

Most of this visual architecture lies at the back of the head, in the occipital lobe. There, over ten billion cells and one trillion synapses create the world we experience. It is here alone, physiology texts state, that visual reality occurs. This is where brightness and color are created and perceived.

So far, so good, until one notices, perhaps idly, that we have just described three different visual worlds. There is the external world, the one in front of us—the realm that we presumably confront or look at. Then there are the upside-down visual images in the retina, formed by those six million cone cells. And finally, there is the third visual kingdom in the brain or mind, where the images are actually constructed and perceived.

Three visual realms. And yet only one appears to us. We don't see double, let alone triple. So which one is *that*? When we now look across our room to a window fifteen feet away, we're entitled to ask: Where is it located? *Where* is the universe?

Language and custom say that it is outside us. That it is "out there." But a smattering of scientists know that this cannot be so. That, in fact, everything occurs strictly within our heads.

The point is ultimately as inarguable as gravity, but its full apprehension requires open-mindedness and scrupulous logic because it contradicts a lifetime of language and custom.

So first, let's really be clear about *where* the visual experience occurs, since this seemingly inconsequential issue has enormous implications. Answer: It's fashioned by those one trillion synapses in the brain. This is a stupendous amount of biological architecture. If you merely tried to tally each of those neural connections devoted to vision at the rate of one per second—not examine them but merely count them—it would require thirty thousand years. This huge amount of physiological structure expends vast energy. And nature, we all know or suspect, does nothing for no reason. So let us not sell it short: The visual realm is perceived in this place alone. There are not multiple visual worlds. There is only one visual kingdom and you perceive it clearly; it is occurring within your skull.

That framed art hanging "over there" across the room is *actually* inside your head. Sure, you always imagined the brain's interior to be dark and mushy, despite reading that complex electric signals and lively energies course there. But now you know what the brain's interior is like. It is there, that framed art, and the window next to it, and the blue sky. All inside the mind. Indeed, even your brain and body are representations in your mind.

But, you may protest, aren't there two worlds? The external "real" world, and then another, separate visual world inside your head? No, there is only one. Where the visual image is perceived is where it actually is. There is nothing outside of perception. How could there be?

"People are so sure that they 'look out' at the world!" says Canadian physicist Roy Bishop, a senior editor of the *Handbook of the Royal Astronomical Society*, never ceasing to be amazed that most folks do not see the obvious. But the illusion of an external world comes from language. Everyone you meet participates in the same charade. It's not malevolent, but useful, as when we say, "Please pass the salt over there." What purpose would it serve to ask for that salt shaker "inside your head"? It is customary to allude to the world as existing outside of us.

"All right," you may say, a bit hesitantly now, "but if that window is within my skull, what about my fingertips that I'm holding up? Don't they define the outer limits of my body?" No, they do not. Those fingers are also within your mind. They are the mind's representation—in tactile form when you experience touch, and visually when you glance at your nails and consider trimming or biting them—and they, too, dwell within the mind. *They are a representation of your body that itself exists within the mind.* The window across the room, and the framed art on the wall, are no farther away than your fingers. They are all equally within the mind.

Of course, we usually define distance as the seeming gap between our mind-bodies and, say, that mind-tree. Our mind-legs require effort and a long interval before we reach the tree that's equally within the mind. So we call that a gap or space or distance, and that's fine, it's how we all express things—as how the mind's body portrayal relates to the other objects in the mind. And, granted, it can take a while to get accustomed to thinking of that stroll as occurring strictly from one part of your mind to another. And that at no point is your mind's representation of your body ever separate from anything else you observe in the world. Yet all this is true.

Colors are created by us. The entire visual universe is located here, not out there. There is no such thing as "out there."

Now, if "that" is within myself, then in a very concrete sense, everything I see is "me." I do not end, not even at the Moon and beyond—at least visually, and aurally, and perceptually.

But can't I at least establish a boundary between self and other in terms of control? Obviously I can clap my hands, but I cannot wiggle your toes. There seems some kind of real, practical demarcation.

Alas, here, too, with the control business, we get into a can of worms. Most people assume they can control stuff, even if their decisions pop up spontaneously. We do not know how we make a decision; it just somehow occurs. We don't know how to make our hearts beat or to perform the liver's five hundred functions. We don't even know how to snap our fingers to music, because if we thought about it, too many muscle and nerve movements are involved and we don't really know how to command them. We just do it. And despite most people (but not Albert Einstein) insisting that they have free will to control their bodies, minds, and lives, much experimental evidence since 1998 shows that this, too, may be illusion. We're not going to "go there" and explore the seeming dichotomy, long debated by scientists and philosophers alike, whether our lives operate via the mechanism of free will, or determinism, or unfold spontaneously, or maybe even by some fourth process we have yet to articulate. What's central here is that the entire house-of-cards separation between me and other, and body interior and exterior, and nature versus ourselves, are relative concepts involving yet more neural connections that impart assumptions about reality.

Reality is an active process that always involves our consciousness. Everything we see and experience is a whirl of information occurring in our minds, shaped by algorithms (represented here by digital zeroes and ones) that create brightness, depth, and a sense of time and space. Even in dreams, our mind can assemble information into a 4D spatio-temporal experience. "Here," said Emerson, "we stand before the secret of the world, there where Being passes into Appearance, and Unity into Variety."

We need to get past them all. We need to see what's baseline, bottom-line real, in our quest for grasping the nature of the cosmos. Doing so, the accurate perception of everything visual as occurring in our mind is perhaps the easiest starting point. That this usually draws blank stares is a function of years of assuming otherwise.

Early in 2015, we asked Dr. Bishop if he could suggest ways to help people "get it." Here are two.

First, light travels from the so-called external world to our eyes. Most people having at least a smattering of science knowledge would surely agree with that. Yet most people believe that they look "out" at the external world. Does not the contradiction of these two ideas suggest that one of them is wrong? Unfortunately our very language reinforces the wrong idea: We say "look in" the cupboard, "look across" the street, "look at" the Moon, "look through" the telescope. Despite acknowledging the direction that light travels, nearly everyone thinks that they look "at" things, that their visual world coincides spatially with an external realm!

Second, that color does not exist external to the observer is more difficult to appreciate, because various color phenomena can be "satisfactorily" explained based upon the four types of light-activated cells in the retina: three cone cells sensitive to red, green, and blue in bright light, and a single type of rod-shaped cell that responds in dim light (i.e., photopic and scotopic vision, respectively). The absence of color in a scene lit by a quarter Moon, color "blindness," contrast phenomena that can generate rich color sensations, and the like can all be accounted for while assuming that retinal cone cells are "color receptors," as if colors were part of the external world. *Not until a person "gets it"—that he does not look "out," that his visual world is a private sensation deep within his brain, that each and every visual scene he experiences resides there*—is it possible for that person to grasp that those indescribable hues are generated there, too. There are obvious evolutionary advantages in having visual spectral discrimination, and our brains evolved a simple way of providing such discrimination: with hue sensations.

It is not necessary to negate the external world. We needn't say that it doesn't exist. It is enough to see through the false assumptions that we "look at" an external world while simultaneously (and equally erroneously) believing that a separate visual world lurks somewhere inside our skull despite it being seemingly imperceptible.

What's important is to grasp that the two-world assumption is illusory. That the world we see *is* the visual perception located in our head.

Language aside, there is no actual "me" performing an act of "looking out." The "me" is a figure of speech corresponding to nothing at all, as vacuous as the word *being* in the phrase "being empty." Rather, everything we see *is* the mind. The silverware on the table might be thought of as being situated in front of us, but its actual location is inside our heads. Indeed, with a little genetic engineering, you could probably make everything that's red move, or make a noise instead, or even make you feel hungry. Or want to have sex—which is what that color can do to some birds. Did tampering with your brain circuits alter an external universe?

Apprehending the cosmos as a single deathless entity synonymous with consciousness may require multiple logical steps, or it can be realized in a single "Eureka!" moment. Like those optical illusions where a set of stairs seems headed downward until suddenly everything changes and it's perceived entirely differently, this reality may have a similarly abrupt onset—a marvelous experience indeed.

This is why so much time is now invested in this vision business—"How many in the world see this?" When asked this very question, Dr. Bishop produced a wonderful reply:

"Have I personally ever met anyone who 'gets it'? I have a friend I have known most of my life. We have discussed many things over the years, including vision. He 'gets it,' as demonstrated in the following text that he wrote a couple of years ago as the caption to a photo of an autumn scene for a calendar produced by a local natural history society:

> This autumn scene presents a feast of color typical of the season. Lightrays (electromagnetic waves of various frequencies) reflect from leaves, . . . [and] are processed by the brain which forms an image within the darkness of the skull. By some feat of mental projection, we have the overpowering impression that the image we experience is located out there beyond our noses. It's a wonderful illusion.

"This isn't rocket science," Dr. Bishop continued. "No math is involved, and minimal science. But what is involved in 'getting it' is a complete break with how one thought vision worked ever since early childhood. Vision operates so flawlessly, so easily, so marvelously, with no effort whatsoever on the part of its owner, that it takes a major leap of insight and introspection to make the transition from the naive assumption that one's visual world coincides spatially with the external world, to the realization that the brightness, detail, colors, and three-dimensionality can only reside somewhere within the absolute darkness of one's skull. That is a big mental leap, which for most people seems impossibly difficult to make. It is so easy to be misled by popular misconceptions. Even amongst scientists, most of whom have never thought much about vision, my guess is that fewer than 10 percent 'get it,' possibly far fewer. The percentage is surely larger amongst perceptual psychologists and physiologists.

"In my own case, I had a Ph.D. in physics before I appreciated where my visual world resided, before I realized that colors and brightness are sensations served up by my brain. That revelation hit me in the autumn of 1969 while reading a small book entitled *The Rays Are Not Coloured* (1967) by W. D. Wright. Wright took the words for his title from Newton's classic book *Opticks* of 1704. Newton was one of the first to 'get it.' The fact that I bought Wright's book and read it indicates that finally, at age thirty, I was ripe for making that leap of insight. All this is but one of the magical aspects of the world, an aspect that helps makes this short life so interesting."

We only ask the reader to let this vision business sink in, to percolate.

"The only things we can ever perceive," said George Berkeley, for whom the campus and city were named, "are our perceptions."

There is no universe without perception. Consciousness and the cosmos are correlative. They are one and the same.

INFORMATION
PLEASE

13

Pure logical thinking cannot yield us any knowledge of
the empirical world; all knowledge of reality starts from
experience and ends in it.
—Albert Einstein, *Ideas and Opinions* (1954)

Reality is a swirl of information in the mind.
    This means that absolutely everything, from the trees
"out there" to our sense of time and perception of distance, is all being continually constructed and perceived by
lightning-quick life-based information systems. Let's examine
how this works.

It's sometimes said that all moving objects, not just sentient creatures, are stirred by information. A falling hailstone
senses the information of a gravitational field and responds
accordingly. By most definitions, information works through an
exchange of energy, so the falling bit of ice is indeed interre-
lating with the field by contributing to the mass of the planet.
More obviously, you yourself invariably gain knowledge via

energy absorption; for example, by capturing a stream of photons like the words on this page arriving on reflected light, or by recognizing meaning in changes of air pressure—the shout of "Hello" from a friend. If information is defined as everything involved in cause-and-effect exchanges, then information interactions are continuous and omnipresent on all levels.

Some people create categories by maintaining that information need not involve a sensory-equipped observer, as when a comet "responds" to the solar wind by pointing its tail away from the Sun. If so, then virtually everything is information, for which individual science disciplines have their own categories and nomenclature systems. Some of these are indeed relevant to consciousness and awareness, if even on an abstract level. But if every possible energy exchange in physics, chemistry, and biology is considered an information encounter, such as the bonding of hydrogen atoms to oxygen to create a water molecule, which occurs in less than a trillionth of a second, then that concept becomes so vague that it's almost limitless what we might characterize as information transfers.

By contrast, if we use the word *knowledge*, then the exchange must involve a sentient organism. But again, because biocentrism maintains that everything lies within consciousness, and nothing exists apart from observers, we'll use *information* in a wide sense.

With all such definitions and qualifications behind us, let's explore conscious animal-based systems and also how current technology intertwines with it, imparting super-high rates of knowledge acquisition—levels that challenge the brain's own architecturally dictated absorption capacities.

Even though consciousness has many deep, fundamental mysteries, it wouldn't be wrong to call it an avalanche of information in the brain, itself an amalgam of the so-called external and the so-called internal coding mechanisms that let the mind create a vast world to make sense of things on multiple levels.

Many of these information algorithms require no learning; they are hard wired even before birth. It's astonishing the complex multitasking actions we animals are capable of solely from what has been genetically programmed. Even plants need no schooling, but automatically respond to wind, gravity, direction of light, water, and various other impetuses, as we'll explore in chapter 15. In any event, the first bedrock issue in information exchange involves methodology: namely, whether the knowledge is gained directly or indirectly. Direct information might be you feeling the Sun's warmth. No symbolic language or intermediary is needed; you straightforwardly sense the solar heat via your nervous system, and its reality is thus inarguable. (Actually, to contrarians *everything* is potentially arguable. In this instance, if we wish to be picky, you really only perceive your skin's atoms jiggling at a faster pace: quicker-moving atoms are what we call heat. The atoms have been stimulated to increased motion by invisible solar infrared light, which humans cannot directly sense. So when we enjoy the Sun's warmth on a spring day, we're actually feeling the speed-up of epidermal atoms caused by an invisible form of light. Nonetheless—it's a direct experience.)

By contrast, all the information you just read was not direct at all. That business about infrared was acquired by you through the use of symbols—words—each of which signified something other than itself; the word *Sun* is not the actual Sun. Such symbolic knowledge is representational and, in contrast with direct knowledge, is subject to revision and possible future improvement. That doesn't mean it's not real. Certainly, actual physical neural connections in your brain were formed, some permanently, after you read the preceding paragraph, particularly if you found it interesting. Moreover, a waiter warning you that an iron sizzler placed on the table is hot carries information every bit as valid as if you had acquired it by inadvertently touching the metal. One method was not superior to the other in terms of knowledge-gaining effectiveness.

A dog barking to alert others in the neighborhood is a good example of secondary information. The other canines inferred meaning in the tone, loudness, frequency, and urgency of the first dog's barks, and instinctively understood that it meant something—something entirely different from the barking sound itself. They took it to mean "a stranger is approaching," and reacted to this information.

Thus, indirect, symbolic information is nothing to belittle. Some forms of it are little short of amazing. Dolphins have the ability to make an extremely complicated series of sounds that implant an image in the minds of other dolphins. They can paint a picture of something of interest—a school of food fish it just spotted, for example—and perhaps even include in the image a kind of italics to highlight areas of emphasis.

We humans use both types of information acquisition, and usually do so without paying much attention to the distinction. As for the physical method of gaining data, the word *analog* only started being used to describe information retrieval and storage methods when we needed to compare it with the new on/off, zero/one, yes/no language of computers and music storage, because those two methods comprise the only choices. Naturally, the labels *analog* and *digital* also arise when we consider the information acquisition, storage, and transmitting architectures of higher-order life forms, those with advanced nervous systems and brains. So, which is it? Does our own operating system (the brain and mind) work digitally or through analog architecture? Much popular literature gets this wrong.

We first need a primer on what those terms mean. Usually analog information systems use waves of some sort, or smooth transitions from one state to another, like a pulse that grows from zero, reaches a certain peak, then subsides. Certainly it is a continuous process. Expressed on a graph it looks like a series of smooth hills with no breaks or pauses. The values it can express are essentially infinite in number because they can be anything at all. Household current in the United States, for example, consists of electricity that pulses 60 times per second, at a nominal 120 volts that vary from plus to minus. In practice the extremes can and do vary and could be, say, −117.77819 volts one minute but 118.9980003 the next, and no one would notice or care; it would still accomplish its task.

In analog technology, a microphone might be used to record pulses of sound (complex air pressure variations) that similarly mutate in a limitless fashion, are translated into varying electrical pulses, and are then recorded on a tape via the rearrangement of tiny, magnetic iron particles for storage. At a later time, this signal can be read, sent to an amplifier and then a speaker, where another magnet causes its cone to pulse at both slow and rapid rates, moving air in the room that reproduces the music. The entire process incorporates a universe of possibilities, and this is analog.

Digital is a different ball of wax. It's rarely used by nature. Gone are the infinite wave possibilities with their myriad nuances. Now, all information has discrete values with nothing in between. In practice, the encoding consists of a series of "on" or "off" signals. This can be accomplished in many ways. With a music CD, about 5 milliwatts of light are employed: monochromatic (a single, narrow color) light works best, and a laser is the perfect device for inexpensively producing and focusing such energy. The light source narrowly aims the light onto the CD's grooves, which contain some four billion minuscule pits that don't reflect light, alternating with flat areas called "lands" that do reflect it to a detector. Each reflection is counted as a yes signal, a one, while a lack of signal means no, or zero.

In practice, the rapidly spinning CD lets 44,100 bits of information be sampled every second in the form of these ones or zeroes with nothing in between. No infinities are to be found here, no limitless possibilities. Instead, the ones and zeroes employ a binary language to create ordinary numerals. With 44,100 numbers per second of music being played, all within tiny channels or grooves (which, if unrolled, would stretch 3½ miles), a rich lode of data is sent to the digital amplifier, which understands what the coded numbers mean and turns them into voltage waves. These surges go to the speakers, which act just as they did before with analog, pulsing appropriately to rapidly disturb the air pressure in the room in the complex fashion we recognize as music. Ultimately, the end result is the same.

So why is digital considered superior by many? Well, waves can get polluted with unwanted noise or degrade with storage, while the ones and zeroes will always be ones and zeroes, and thus tend to be far more immune to distortion or loss over time. Moreover, clever algorithms that look for patterns in the numbers can compress them so that they take up less storage space. You can't do that with waves.

When it comes to the brain's functioning, it's natural to imagine that its operations are purely digital, too. On the cellular level we'd assume that a neuron either fires a signal—sends a pulse of electricity—or it doesn't. This would seem to precisely define a digital operating system. Moreover, because digital is all the geeky rage these days, it's natural to imagine that our ultra-sophisticated brains must surely operate using the latest and greatest technology. But in real life, wouldn't you know it, a brain is far more complicated than that. (If you're enjoying learning all this, it's because the brain generally likes reading about itself.)

First, each neuron achieves its goal of stimulating or communicating with another (or several other) neurons not by merely "pulling the trigger" once, but rather by a *series* of electrical firings. It can change its signaling intensity as well as the rapidity. A more rapid series means a stronger signal. Such variations produce complexities far beyond a mere zero-or-one state of affairs, but rather denotes a system whereby the brain's nerve cell signals are ratcheted up or tuned down, along with frequencies that amount to a continuum, which means the brain is an analog machine.

And there's more to its complexity than even these signaling subtleties. A neuron typically receives electrical indicators from

several others, and some of the incoming signals can be excit-
atory, whereas others cause suppression. The entire cascade is
like a symphony where the individual instruments modulate
their strengths in complex ways. So what a particular neuron
"decides"—its ultimate output—is the result of the sum of all
the varied signals it is receiving, which definitely lies along a
continuum and therefore is not at all digital.

Moreover, not only does the frequency or power of electrical firings change, but physical neuron connections with their
neighbors vary in their strength, and this, too, lies along a wide
range that is anything but a yes-or-no situation. A neuron can
have more than one synapse (connection point) and it can be
distant or close to the main body of the nerve cell (which matters), or else forms a tight bundle with many others, or comprises
a sparser, outlying connection. With so many possibilities even
within the tiniest sample of brain tissue, the aggregate of all the
ways signaling can unfold is staggering. Expressing the possible
different brain connections would require a number depicted by
a one followed by more zeroes than could fill every line of every
page of this entire book. It's not much of an exaggeration to call
the brain's/mind's potential, or its capability for variety, limitless.

What's cool is when we make our own latest technologies
form interplays with our minds.

Say we want to experience a movie, even one that's in 3D.
Let's do it.

Not long ago this technology involved an analog process
using film, where every spot on each frame could receive any of
a continuum of brightnesses or colors. Moreover, the early years
of motion pictures taught us that the original frame rate—16
per second—lay within the mind's "flicker fusion threshold"
of 20 flashes per second. That is, showing 16 different images
per second, with a moment of darkness in between, as films
did during the silent movie era, was insufficient to prevent the
mind from seeing separate bits of darkness. Everyone perceived
a flickering.

The advent of sound also brought the introduction of a major visual cinema improvement. Because our minds "remember" and thus merge together images that arrive faster than about 20 per second, movies abruptly went to 72 images per second, which created seamless motion with no trace of flicker or pulsations. In practice there's really only 24 *different* images per second in motion pictures, but each frame is shown three times before the next image appears three times. The point is that our technology always has to be designed to operate in tune with the vagaries of our mind's architecture, including its quirks.

Film worked well, but nowadays, with enough capacity, each part of a digital camera's charge-coupled device (CCD) chips encodes enough of this same information in binary fashion that the result need not be inferior. Nonetheless, image quality *is* inferior to 35 mm film, even in theaters using the newer industry-standard 4K projectors—and it's downright blurry in the many movie houses that still have 2K projectors. However, when the same 4K image, composed of about eight million individual pixels, is less enlarged by being confined to the size of a home TV, even one with an 80-inch screen, the picture seen at normal distances exceeds the resolution threshold of the eye-brain visual system, and is gorgeous in its detail.

If the movie is encoded on a DVD, where fifty gigabytes of data can be utilized for a single Blu-ray movie, the 3D effect merely requires that each eye see a different image. In the '50s this was accomplished with black-and-white film that had simultaneous blue and red versions on it, and the viewer wore red/blue or red/green glasses so that each eye could see one image but never the other. Today's method either employs glasses whose left and right eyes receive individual vertical and horizontal polarization, or else rapidly flickering shutters so each eye alternates in sync with the double images meant for it alone. That these methods create a true 3D sense is instructive about reality.

Anyone with normal binocular vision experiences the wonderful sensation of depth in their visual world. That powerful experience of the third dimension was generated by two-dimensional "stereo pairs" way back in the nineteenth century with the then-popular stereo viewer, and nowadays at an IMAX theater presenting a 3D movie. In all these situations, the two 2D images contain parallax information, meaning the images are subtly dissimilar, just as each eye receives a slightly different image, with nearby objects shifted the most, thanks to the fact that each eye is gazing at things from a somewhat different angle. Yet the observer experiences marvelous depth just as fully as if the actual three-dimensional scene were present in front of him or her. Our takeaway from this: The magical sensation of depth *must* arise internally, when the visual input with its parallax discrepancies is sorted out and presented to the conscious level of the brain. It follows that the rest of one's perceived visual world must be located there, too—not "out there," beyond our bodies somewhere.

It bears repeating: There is nothing "out there" beyond the reality constructed in our minds. Or, if so, it would be utterly mysterious and unexperienced—certainly not the world with its scurrying cars and trees swaying in the wind. All we know and *can* know is contained within our mind/the information processed in our brains.

If this seems impossible to accept, remember that if there were to be some precursor to the colors, brightness, and 3D depth of the visual world we continually enjoy, some "exterior" stimulus, it would be no more than invisibly blank magnetic and electrical fields, since that's what light really is.

Apprehending reality is an ongoing goal-less informational process. But attempting to logically conceive of it is a different project, a piecemeal enterprise. Certainly, no single mental image can adequately capture *Being*. A punch line or single phrase that might fully express ultimate knowledge will remain elusive.

But a good start is simply to see conscious experience as a swirl of information, while abandoning the notion that anything is truly external.

MACHINES WITH AWARENESS

14

Maybe the only significant difference between a really smart simulation and a human being was the noise they made when you punched them.
—Terry Pratchett, *The Long Earth* (2012)

The famous physicist made headlines at the end of 2014. Stephen Hawking, in a BBC interview, spoke about how we should be very wary of developing "full artificial intelligence" (AI) as it "could spell the end of the human race."

His doomsday musings were hardly original. SpaceX's Elon Musk had said the same thing earlier that year, warning that AI is "potentially more dangerous than nukes." The worrisome idea of computers possessing greater-than-human intelligence coupled with a sudden independent consciousness was first termed "the Singularity" back in 1993, in a paper by computer scientist and writer Vernor Vinge. And although his initial predictions about vast computer improvements merely mirrored the foresight of others—like the expected frequent doubling in computer power envisioned by Intel cofounder Gordon Moore in 1965—Vinge believed it would lead to "change comparable to the rise of human life on Earth."

As we all know, computers already control and facilitate much of our daily lives from banking to robotic automobile assembly, and no one wants to return to the old days of manual drudgery for menial tasks like repetitive spot welding. We're even used to machines understanding commands and correctly responding to questions. Major advances are reported annually. In 2015, a team in Berkeley, California, unveiled a new, powerful AI technique—a "deep learning" architecture—that lets a robot quickly learn new tasks with only a small amount of training. That robot rapidly learned to screw the cap on a bottle, even figuring out the need to first apply a slight backward twist to find the thread before turning it the correct way.

The fear generated by the Singularitarians is that artificial intelligence will someday reach a point of complexity where *the machines become self-aware*. It is this trait that produces the sci-fi fantasies of machines designing better robots and computers for their own purposes, and in a way that bypasses human control.

We have of course seen this theme in films like the Terminator series, *Westworld* (where a gunslinging robot runs amok at a theme park), and in *2001: A Space Odyssey*. But there's a very clear and spooky distinction that arises with the Singularity. It is one thing for computers to screw up in some fashion that causes us trouble. It is quite another for them to gain perception.

The creepy self-aware business is given credence because it's promulgated by a few reputable authorities, like Cornell University computer engineer Hod Lipson. He's pointed out that with ever-growing complexity, computer problems will increasingly require that we design them to deal with split-second issues by adapting and making decisions on their own. As machines get better at learning how to learn, Lipson believes it invariably "leads down the path to consciousness and self-awareness."

This brings up a new issue, an important one: What is the basis of consciousness? If complex electrical circuitry plays a key role, well, computers are obviously getting that. Would we even be able to recognize sentience in a machine? Already, researchers at Yale University have created a robot named Nico that can recognize itself in a mirror, and make decisions about spatial recognition based on its own position and its surroundings. It even knows when an object is merely a reflection in a mirror, rather than naively thinking that it exists behind the glass. Nico's creators and programmers speak of machines "autonomously learning about their bodies and senses."

With supercomputers improving their capabilities, and speeds of 4 exaFLOP/s (or 4×10^{18} calculations per second) expected by 2020, might we actually arrive at the Singularity—the amazing event predicted by Vinge and seconded by people like futurist Raymond Kurzweil, the man who designed the first text-to-speech synthesizer? In Kurzweil's 2005 book, *The Singularity Is Near: When Humans Transcend Biology*, he flat-out predicted that by 2045 the first computer will become self-aware. After the arrival of this dreaded Singularity, humans and animals will share Earth with another intelligence, possibly forever.

Needless to say, all this catches the notice of everyone who knows that the external world and consciousness are linked if not correlative. So as we read these predictions of machine sentience, a little skepticism may be realistic. We've never seen inanimate material suddenly come to life. Even if future computer brains are designed to more closely match the architecture of ours, why should that bring the silicon entity to true self-awareness? As Dutch computer scientist Edsger W. Dijkstra said, after winning the 1972 A. M. Turing Award, "The question of whether a computer can think is no more interesting than the question of whether a submarine can swim."

After all, there is a huge functional gap between the human and the computer mind, and even comparing performance levels is an apples-and-oranges affair. Computers possess vast search engines that can call up data with an efficiency far beyond what human brains can accomplish. But computers fail when it comes to most stone-simple human tasks, like understanding nested structures in the language of someone trying to convey subtle concepts, or creating ideas based on hierarchical symbols that then form higher-order ideas.

But again, lurking behind any capability comparisons is the bedrock issue of what is involved in an entity being conscious. It's easy to assume that if consciousness is generated by an electrical current that stimulates appropriate neural inputs, well, machines use electrical circuits, so are we halfway home?

Research into consciousness has been ongoing in Europe and the United States. In 2014, European researchers reported results (in the journal *Nature Neuroscience*) of their investigations into how "higher-order consciousness"—abstract thinking and reflexivity—is generated by electrical currents called gamma waves. The researchers fired low-voltage currents through test subjects' frontal lobes to mimic the gamma band in an effort to induce self-awareness in unconscious patients. It worked. The dreams experienced by the test subjects started to become lucid. The researchers concluded that conscious awareness is induced at electrical currents pulsing at 40 cycles per second. It all strongly implies that the subjective experience arises at least in part because of electrical stimulation.

The mapping of brain activity has been under way for decades. Yet its usefulness in understanding consciousness remains a topic of heated debate, partly because brain activity is often scattered throughout that organ, and in variable formations, and partly because learning what regions control which functions and feelings is not the same thing as understanding what's really going on when we experience sensations.

As Dutch researcher José van Dijck explained in a 2004 article, "Memory Matters in the Digital Age," "The brain is less like a computer and more like a symphony; it continually plays variations on a theme when it comes to activities like recalling memory. Even if we can track brain activity, we can't describe the processes that occur."

The issue of consciousness often wiggles away from easy understanding. The public seems generally oblivious that it harbors any sort of deep mystery at all. Some regard awareness as a mere ancillary property of life, a casual characteristic that evolution produced to give complex life forms an advantage. Many people seem unaware that—whether discussing the possibility of computer Singularity, or our own experiences that are so central in this book—consciousness is a profound issue. Without any exaggeration, it may well be characterized—as Paul Hoffman, former *Encyclopedia Britannica* publisher, told one of the authors—as the deepest and most important in all of science.

It has certainly plagued scientists and thinkers through the ages. In a letter to German theologian Henry Oldenburg, Isaac Newton wrote: "To determine . . . by what modes or actions [light] produceth in our minds the Phantasms of Colours, is not so easie." And nineteenth-century biologist Thomas Henry Huxley, one of Darwin's early advocates, called the state of consciousness "remarkable" and said that it "is just as unaccountable as the appearance of the Djin when Aladdin rubbed his lamp."

In modern times, researchers are often hotly divided even when defining awareness. Tufts University philosopher and cognitive scientist Daniel Dennett probably started the modern embroilment with his five hundred-page 1991 book, *Consciousness Explained.* Since the work was overwhelmingly involved in such issues as describing which areas of the brain are associated with which functions, and only at the end contained a brief concession that consciousness (if defined as experiencing things) was a total mystery, critics howled and still do so. Several alluded to the work as "Consciousness Ignored." We should at least be able to articulate what it is we are trying to grasp. Yet even this is easier said than done. Writing about consciousness, Stanford physicist James Trefil says that "it is the only major question in the sciences that we don't even know how to ask."

Yet, mysterious or not, we surely want to know whether awareness is amenable to being described in physical terms, such as the sum total of neural processes in the brain. If, in the fullness of time, we find that consciousness *cannot* be explained exclusively by physical events, then, possibly like the mysterious vacuum energy that fills the cosmos, it may require an explanation using nonphysical means. This sounds uncomfortably on the precipice of magic. Nonetheless, some philosophers resolutely maintain that consciousness is indeed nonphysical in nature. But if so, then what outside of the physical or biological sciences is required to explain it? Or, like love and other imponderables, must it remain inexplicable?

Modern research revolves around brain function, and although a handful of researchers have claimed that such patterns "explain" consciousness, very few in the field agree. To at least set the issue into a more digestible outline form, Australian philosopher and cognitive scientist David Chalmers has divided the topic into "the hard problem of consciousness" and "easy problems" such as "explaining the ability to discriminate, categorize, and react to environmental stimuli." Also in the easy camp are projects attempting to map which parts of the brain are linked with which sensations and functions.

Easy problems merely require that researchers learn the biological or neural mechanism that can perform various functions. We can already see that these are potentially soluble, perhaps completely, and that mapping the brain can be entirely consistent with what we know of natural phenomena.

The hard problem, which is actually a very simple issue despite its eluding general public awareness, is explaining how and why we have subjective experiences at all, such as seeing and hearing. Somehow (according to science's mainstream view) the inanimate materials that comprise our bodies—carbon and minerals and electrical pulses—find a way to bestow on us the experience of *feelings*.

Now, it's assumed that the Sun has no feelings. And rocks cannot "enjoy" the warm sunlight striking their surfaces. Yet *we* savor the smell of fresh-cut grass, feel pain if pinched, experience thoughts, and sense the rich crimson of a sunset. We *feel*. How and why? It's the most basic kind of question, yet it has no answer to date.

The depth and profundity of this goes to the heart of biocentrism, and to the paranoia over possible computer Singularities, and to the very quest to apprehend the cosmos. Nothing escapes the sweaty grip of *perception*. We need to know what it is.

This is the issue obsessing a new breed of researcher, like University of Southampton computer researcher Stevan Harnad. When answering questions about the work of putative consciousness investigators like Dennett, he is ruthless in cutting to the central issue and sloughing off all extraneous hand-waving.

The right question is not about our thoughts or memories, or what may or may not be illusory, he says. "Thoughts" is 100% equivocal. If it just means "internal goings-on that generate certain outputs in response to certain inputs," then no problem (and no problem solved!). But if "thoughts" means "*felt* thoughts," then you might as well call them "feelings" (what it feels-like to think and reason is just one instance of the multi-qualitative world of feelings; there's also what it feels-like to see, touch, want, will, etc.) . . . The persistent niggler, though, is how and why all that admirable hierarchical Turing function should be *felt*.

This becomes the bottom line. Why and how do we *feel*? How does this sense of perception, or awareness, or consciousness arise? What is it, really?

This goes to the heart of everything. We do not know how consciousness arises in individuals at birth. Some Hindus believe the soul or individual sense of self enters a fetus at three months. But how can they know? Is there really anything that enters anything else? We all recognize this sense of awareness;

it's more intimate than anything else. It intuitively feels like it's beyond time and space (which in fact it is). Our memories are limited and selective, but consciousness has always been our deepest companion. Our true self, truth be told. Does it arise at all? Is it eternal? Nobel Laureate and physicist Steven Weinberg was far from alone when he conceded that there's a problem with consciousness: Its existence doesn't seem to be derivable from physical laws.

Biocentrism shows that our sense of external/internal is a mental classification scheme, wherein all sensations are here and nowhere else. Nothing is truly external, meaning outside the mind. We may believe this consciousness has a home in our brains, and there's a relative truth to that, but not an absolute one, because the brain itself is as much a construction in our minds as the supposedly external trees and tablecloths.

The brain? Sure, we've seen films of autopsies and assume that the mushy three-pound blob is where it all happens. But what is that brain, really? We, unlike Zeno of Elea, assume innumerable separate things exist in our universe, and that the brain is one of them and our consciousness exists within the brain. Objects and more objects.

But what's really there? Energy fields exist everywhere, and the solid things we see and touch are merely artifacts of our selective sensory architecture. If our sensing algorithms had been differently structured, we'd see nothing on the planet, because its true nature is essentially emptiness accompanied by omnipresent invisible energy fields. Yet that isn't "true," either, because nothing is, until it's perceived.

What we do observe, all this richness, is a deliberate spatio-temporal algorithm attuned to particular electromagnetic frequencies. Push your finger down on the table top and it feels solid. But no solids are ever contacted, not for an instant. Rather, the outermost atoms of your skin are surrounded by negatively charged electrons, and these are repelled by the similar electrons in the table. The sense of solidity is illusory; you feel only repulsive electrical fields. Fields. Energies. Nothing solid, ever. And it all occurs within the detector (mind), which imparts a sense of space (location) and time, which otherwise have no inherent reality. Indeed, the universe can be viewed as a blurry, probabilistic state of potential information, which the mind-system "collapses" into actual information and sensations when processed by the mind-system. It's a unitary process that bestows the feeling of a "me"—the sense of being.

Yet we see periodic articles in science publications proposing the latest "test" of whether a computer has acquired consciousness. All are clever, but none to date seem foolproof or, even, probably, valid. You know you are conscious because you experience your own awareness. You'd undoubtedly give good wagering odds that other people as well as at least the "higher-functioning mammals" such as dolphins and orangutans are conscious because they're similarly composed, with behaviors not too dissimilar from yours, plus they arose from branches of the same evolutionary tree.

In a 2011 *Scientific American* article, "A Test for Consciousness," the authors, Christof Koch and Giulio Tononi, propose such tests to probe for actual awareness. Say these researchers: "How would we know if a machine had taken on this seemingly ineffable quality of conscious awareness? Our strategy relies on the knowledge that only a conscious machine can demonstrate a subjective understanding of whether a scene depicted in some ordinary photograph is 'right' or 'wrong.' This ability to assemble a set of facts into a picture of reality that makes eminent sense—or know, say, that an elephant should not be perched on top of the Eiffel Tower—defines an essential property of the conscious mind. A roomful of IBM supercomputers, in contrast, still cannot fathom what makes sense in a scene."

But, again, producing correct responses to such tests seems off-point and irrelevant for establishing consciousness. Critics quickly wrote letters that the magazine published, arguing that "a human child under a certain age would not be able to pass this test, nor would an adult in a dream state, or under the influence of hallucinogenic drugs. Yet nobody would deny that these humans [have consciousness]."

In short, we ourselves must first learn to recognize the difference between a tough computational problem and actual perception. Again, it probably comes down to *feelings*. Can a computer feel things like pleasure and pain?

In the ordinary way of conceiving things, we regard consciousness as having individual centers—you and me and each raccoon. We imagine it arises at birth and subsides at death. Since it thus seemingly comes and goes, the issue of whether it can arise in a machine certainly seems reasonable.

But if consciousness is correlative with the cosmos, then the question defaults to an inquiry into the *entirety of existence*. Tackling this is the same task as pondering the overarching universe. Though a valid and venerable topic, all methodologies employing symbolism limited to representing individual parts must fall short. Logic and comprehension—which always employ symbolic language—would only be useful if those representational "parts" effectively convey a new meaning about the whole.

They don't. They can't. And this reveals why cosmology's attempts to "explain" the universe have always seemed bewildering and incomplete. No answer satisfies, partly because our questions are trivial and inconsequential. This can't be avoided. We think and speak using language, which in turn employs words that are all symbols for something else. This is an adequate process for engineering bridges or asking someone to pass the mustard. But it fails as soon as it involves something beyond symbolism such as ecstasy, love, certain empathic feelings, and certainly The Whole Shebang.

Until we understand the nature of space, time, and reality itself, and their biocentric underpinnings, machine sentience will not and cannot happen.

GO GREEN

15

The real Temple is the whole world, and there is nothing as divinely blessed as a blooming growing garden.
—Vera Nazarian, *Dreams of the Compass Rose* (2002)

n the popular movie *Avatar*, humans mine a lush moon inhabited by blue-skinned extraterrestrials, the Na'vi, who live in harmony with nature. Human military forces destroy their habitat despite objections that it could affect the bio-network connecting its organisms. On the eve of the big battle, the protagonist, Jake, communicates via a neural connection with the Tree of Souls, which intercedes on behalf of the Na'vi.

We think of time and consciousness in human terms. But like us, plants possess receptors, microtubules, and sophisticated intercellular systems that likely facilitate a degree of spatio-temporal consciousness. The movie suggests that we don't understand the conscious nature of the life that surrounds us.

"Although I saw the movie three times," one of the authors (Lanza) wrote in a *Huffington Post* blog, "I still cringe whenever someone tells me that a plant has consciousness. As a biologist, I can accept that consciousness exists in cats, dogs, and other animals with sophisticated brains. Studies show that dogs have a level of intelligence—and consciousness—on par with a two- or three-year-old human child. In fact, in 1981, Harvard psychologist B. F. Skinner and I published a paper in the journal *Science* showing that even pigeons were capable of certain aspects of self-awareness. But a plant or a tree? To consider the possibility seemed absurd—until the other day.

"My kitchen merges into a conservatory, a mini-rainforest with palms and ferns. While having breakfast, I looked up at one of my prize specimens, a queen sago tree. For the last several months I'd been watching it send up new fronds, which, since the winter solstice, have been repositioning themselves towards the shifting Sun. During that time I also watched it respond to an injury to its trunk by sending out air-roots in search of new soil to re-root itself. It was a clever life-form, but clearly not conscious in any known biological way.

"Then I remembered the episode of *Star Trek* called 'Wink of an Eye.' In this episode, Captain Kirk beams down to a planet and finds a beautiful but empty metropolis. The only trace of life is the mysterious buzzing of unseen insects. When he returns to the ship, the crew continues to hear the same strange buzzing sound. Suddenly, Kirk notices that the movements of the crew slow down to a stop, as if time itself were being manipulated. A beautiful woman appears and explains to Kirk that the bridge crew hasn't slowed down, but rather, he has been sped up, having been matched to the Scalosians' 'hyper-accelerated' physical existence. Back in real time, Spock and Dr. McCoy figure out that the strange buzzing is the hyper-accelerated conversations of aliens that exist outside normal physics.

"We think of time—and thus consciousness—in human terms. In my mind, I could easily accelerate the plant's behavior like a botanist does with time-lapse photography. The feathery creature, there in my conservatory, responded to the environment much like a primitive invertebrate. But there was more to it than that. We think time is an object, an invisible matrix that ticks away regardless of whether there are any objects or life. Not so, says biocentrism. Time isn't an object or thing; it's a *biological* concept, the way life relates to physical reality. It only exists relative to the observer.

"Consider your own consciousness. Without your eyes, ears, or other sense organs, you would still be able to experience consciousness, albeit in a radically different form. Even without thoughts, you would still be conscious, although the image of a person or tree would have no meaning. Indeed, you wouldn't be able to discern objects from each other, but rather would visually experience the world as a kaleidoscope of changing colors.

"Like us, plants possess receptors, microtubules, and sophisticated intercellular systems that likely facilitate a degree of spatio-temporal consciousness. Instead of generating a pattern of colors, the particles of light bouncing off a plant produce a pattern of energy molecules—sugar—in the chlorophyll in its stems and leaves. Light-stimulating chemical reactions in one leaf cause a chain reaction of signals to the entire organism via vascular bundles.

"Neurobiologists have discovered that plants also have rudimentary neural nets and the capacity for primary perceptions. Indeed, the sundew plant (*Drosera*) will grasp at a fly with incredible accuracy—much better than you can do with a fly-swatter. Some plants even know when ants are coming towards them to steal their nectar and have mechanisms to close up when they approach. Scientists at Cornell University discovered that when a hornworm starts eating sagebrush (*Artemisia tridentata*), the wounded plant will send out a blast of scent that warns surrounding plants—in the case of the study, wild tobacco (*Nicotiana attenuata*)—that trouble is on its way. Those plants, in turn, prepare chemical defenses that send the hungry critters in the opposite direction. Andre Kessler, the lead researcher, called this 'priming its defense response.' 'This could be a crucial mechanism of plant-plant communication,' he said.

"As I sat in the kitchen that day, the early-morning sun slanted down through the skylights, throwing the entire room into gleaming brightness. The queen sago tree and I were both 'happy' the sun was out."

The author's turnaround in his appraisal of our chlorophyllic companions, and the idea that we may have previously limited ourselves in what we've allowed into the "conscious life" fraternity, has been gaining scientific respectability for years, starting before the birth of the twenty-first century. The subject has been widely popularized by the likes of UC Berkeley journalism professor Michael Pollan, who has written books and a *New Yorker* magazine article about how plant science is increasingly pointing to a high degree of botanical intelligence.

All this is a bit of a resurrection of the widespread hippie idea of the 1960s, that plants respond if you talk to them and *really* like it if you soothe them with music or pet them like puppies. When the environmental movement burgeoned in the decades to follow, and forests started to be seen as more than mere unprocessed lumber, such plant-kingdom spokes-mammals were pejoratively called "tree huggers."

It all gave way to a new field of science sometimes called *plant neurobiology*, which starts off a bit controversially because not even the most ardent plant-boosters claim that plants have neurons (nerve cells)—let alone actual brains.

"They have analogous structures," Pollan explained in an interview broadcast on Public Radio International. "They [take] . . . the sensory data they gather in their everyday lives . . . integrate it, and then behave in an appropriate way in response. And they do this without brains, which, in a way, is what's incredible about it, because we automatically assume you need a brain to process information."

Apparently, then, neurons aren't necessary in order to have cell-to-cell communication—or even information processing and storage! In a 2012 *Scientific American* article titled, "Do Plants Think?" Israeli botanist and research scientist Daniel Chamovitz insisted that plants see, feel, smell—and remember. But how is this possible without neurons?

Explained Chamovitz,

> even in animals, not all information is processed or stored only in the brain. The brain is dominant in higher-order processing in more complex animals, but not in simple ones. Different parts of the plant . . . [exchange] information on cellular, physiological, and environmental states. For example root growth is dependent on a hormonal signal that's generated in the tips of shoots . . . [while] leaves send signals to the tip of the shoot telling them to start making flowers. In this way, if you really want to do some major hand waving, *the entire plant is analogous to the brain.* But while plants don't have neurons, plants both produce and are affected by neuroactive chemicals!

The most inarguable analogy involves the glutamate receptor, a neuroreceptor in the human brain necessary for memory formation and learning. Plants do have glutamate receptors, said Chamovitz, and "from studying these proteins in plants, scientists have learned how glutamate receptors mediate communication from cell to cell."

But what about experience? Cognition? Consciousness? The experience of sounds? We naturally assume you cannot hear anything without ears. But according to Pollan's radio interview, researchers have played a recording to plants "of a caterpillar munching on a leaf—and the plants react. They begin to secrete defensive chemicals."

Pollan and others claim that plants possess all the human senses and also some additional ones. In a way there's logic to this. In a time-based scheme (which, remember, is only how we perceive things, and shouldn't be construed as existing in an absolute way), plants were on Earth hundreds of millions of years ahead of us mammals. One way of logically interpreting this is that humans have improved on plants—we are the evolutionary branch that grew farthest away from them. But one might alternatively argue that because plants had so much leisure opportunity to improve if they needed to, they already would have done so if there were any benefit in it. By this reasoning, their ancient presence argues for an actual biological superiority, at least on some levels.

No one doubts that all plants can sense the presence of water or the direction of up and down (in other words, gravity), and may even be able to sense an increase of density in the soil ahead of their roots, so they're aware of a potential obstruction before wasting their time and energy having to make actual physical contact with a rock.

Plants even have a memory. And not just some simple reflex where a certain stimulus creates an automatic response. "Plants definitely have several different forms of memory, just like people do," said Chamovitz. "They have short term memory, immune memory, and even transgenerational memory! I know this is a hard concept to grasp for some people, but if memory entails forming the memory (encoding information), retaining the memory (storing information), and recalling the memory (retrieving information), then plants definitely remember."

When we explore the nature/observer correlate, we naturally hold ourselves as humans as the epitome of conscious intelligence. Most of us would include other mammals as well, especially cats, dogs, bunnies, and other favorite people-companions and pets. But is this bias born solely of their familiarity—the fact that we can recognize a face in a way we do not perceive when watching, say, a worm? Or do we instead regard ownership of a brain a prerequisite for joining the fraternity, and only let those with sophisticated neural architecture into the club?

Time is relative to the observer, and despite our human preconceptions, lower animals—and even plants—may experience consciousness albeit in a considerably different fashion from us. Space and time relationships depend on the entirety of the detector, even if that logic is diffuse and not concentrated into a brain-like structure. Plants clearly have a different information and archiving process from the brain, but time is relative to the observer and need not operate on our human timescale. Time is *bio*-logical—completely subjective and invariably emergent from a unitary co-relative process. All knowledge amounts to relationships of information, with the observer alone imparting spatio-temporal meaning. Because time doesn't actually exist outside of perception, there is no experiential "after death" even for a plant, except the death of its physical structure in our "now." You can't say the plant or animal observer comes or goes or dies, since these are merely temporal concepts.

People have long wondered whether plants "feel," even though it's obvious they're very aware of things like gravity, water sources, and light. It's also obvious that they accomplish these perceptions in very different ways from us mammals or even so-called lower life forms. Tadpoles and other amphibians detect light with pigmented cells in their skin so they can adapt their camouflage to different backgrounds; sparrows can adjust their circadian rhythms without using their eyes at all. They can sense light through feathers, skin, and bone! And mice can do the same thing even when blind. So can at least one species of octopus, which even bypasses any involvement by the brain or nervous system, as discovered only in 2015.

Some creatures "feel" rather than "see" light. For example, octopuses and tadpoles (and other amphibians) detect light with pigmented cells in their skin so they can adapt their camouflage to different backgrounds; sparrows can adjust their circadian rhythms without using their eyes (they can sense light through feathers, skin, and bone); and mice can do the same thing even when blind.

One of the mechanisms for sensing light without eyes appears to be a substance called melanopsin, first discovered in 1998 in the skin of frogs. It allows mammals to detect light beyond and separate from the retina's rod and cone cells. This photopigment reveals a previously unknown and primitive nonvisual photoreceptive system.

Sensing light without possessing eyes is an extremely important ability for biological rhythms and may provide us clues when we muse about how plants, which don't have eyes to see the diurnal day/night cycle that long ago established our own biorhythms, might experience the passage of time.

Of course, plants are utterly light dependent, and utilize it via chlorophyll, a molecule that particularly "likes" blue light and can also feast on red, but has no use for green wavelengths. This explains why leaves and grass appear green: We see the part of the Sun's spectrum that has been rejected by the plant and reflected away—not absorbed and utilized. Thus, a tree's leaves are green not because chlorophyll is fond of green, but because it's so indifferent to it that it makes those light photons bounce away.

In any event, life forms lacking eyes, such as plants, obviously rely exclusively on other kinds of sensory methods to experience reality. How they perceive time in the world involves sensing and responding to light in a nonvisual way, probably augmented by other methodologies that skip the electromagnetic spectrum altogether.

In higher-order animals, the brain keeps track of time by creating what scientists long assumed to be a biological version of a stopwatch. But our brains actually operate differently, and are not at all like our familiar clocks, which may also help explain why we can perceive temporal intervals incorrectly as we manufacture the illusion of a seamless flow of reality—the result of our continually inscribing events in our memory circuits. A plant doesn't have a brain, so information and "memories" must be stored in other ways—perhaps in the same way a plant knows in what direction it should grow.

How humans record our sensations of time is still basically mysterious. So it will be even harder to figure out how plants "stretch and twist" all this information to serve their survival needs. Because the passage of "time," in the final analysis, is just a tool organisms create and utilize to perceive what's happening around them and to effectively respond to the flow of their physical environment, plants have obviously done a good enough job at it to survive for 700 million years.

Usually we only call something sentient if it talks or responds to us on the biologic timescale we use. But we may have much to learn about the nature of life from the world of the fictional Na'vi, where the plants have an exaggerated sense of touch sensitivity and can communicate through "signal transduction."

"The plants in the film are fake," says Jodie Holt, a plant physiologist from the University of California, Riverside, "but the science is real."

THE QUEST FOR A THEORY OF EVERYTHING

16

It is not enough for theory to describe and analyse, it must itself be an event in the universe it describes.
—Jean Baudrillard, *The Ecstasy of Communication* (1987)

Science has an obsession: to create a Grand Unified Theory. The motive is noble and goes back centuries. Indeed, the quest originates as something deep inside us all. For, science's purpose is to understand this world and our place in this universe and how things fit together. The more that the universe's forces, energies, phenomena, and structure can be incorporated into an inarguable single matrix, the closer we are to figuring out what the heck all of this is, and what it means.

In the mid-nineteenth century, brilliant thinkers started to show that seemingly disparate phenomena were two sides of the same coin. Specifically, in 1865, Scottish physicist James Clerk Maxwell published his groundbreaking unifying theory of electricity and magnetism. It introduced the correct idea that electromagnetism—now known to be one of the four fundamental forces in nature—interrelates both. After all, the movement of electric charge gives rise to magnetism, as when an electric current in a wire deflects a compass needle.

In a flurry of exhilarating revelations soon after the birth of the twentieth century, we learned that matter and energy are likewise a single essence and convert from one to the other. This was a greater leap because a piece of chalk and a flash of light really do seem like totally dissimilar entities. That they are of a single essence—and that we've actually come to observe energy change to matter (e.g., the creation of matter and antimatter by the collision of high-energy—gamma ray—photons) and matter to energy (in H-bomb explosions)—remains more than revelatory for most of us; it is amazing.

The endpoint seemed clear: Perhaps *everything* is a single entity and we need only find how exactly the four fundamental forces and the three fundamental particles interrelate, and how they got that way. The fundamental forces are gravity; electromagnetism, which includes electric and magnetic fields; and the strong and weak forces, whose ranges are minuscule so that they operate solely within atoms. As for permanent fundamental particles, one can start with the quarks that bind together in threesomes to form the nucleons of every atom, plus electrons, plus neutrinos, which are the most prevalent yet do not bind together to form structures. However, beyond these, today's standard model gets complex, with antiparticles, and those like muons with short life spans, and even "force-carrying particles," so that listing an absolute figure for the number of elementary particles is more complicated than it sounds.

Einstein, whose relativity theories of 1905 and 1915 nailed the mass/energy business, spent the rest of his life hunting unsuccessfully for a Grand Unified Theory that would unite these with everything else, including the most elusive: gravity. Meanwhile, quantum mechanics showed how objects behave in the sub-microscopic realm, and this ignited a race to unify quantum mechanics with relativity, a quest that continues to this day.

It was a heady century, from the mid-nineteenth to the mid-twentieth centuries. But then further progress in fundamental physics pretty much ground to a halt. Brilliant minds kept hunting for the Holy Grail but it was nowhere to be found. In the closing decades of the twentieth century came string or superstring theory, or M-theory. Several physicists, using advanced mathematical ideas, showed that at least three of the four forces could emerge if, in the realm of the super-tiny, a trillion trillion trillion trillion trillion times smaller than an atomic nucleus, reality consisted of one-dimensional strings. Depending on how they connected or looped, the basics of the universe could be created.

Except, it didn't work. Not in our reality anyway. As discussed in chapter 10, to make it happen on paper there needed to be eight new dimensions, each with specific mathematical properties. The problem with this was obvious from the start: There is no hint that any of these dimensions actually exist. Neither our senses nor our instruments suggest their reality. Worse, if they did exist, no observation or experiment could possibly detect any of them. Thus, string theory is nonfalsifiable— you can't devise a test that can prove it right or wrong.

All right, maybe the individual dimensions cannot be tested. But perhaps string theory's overall thesis can come up with a testable prediction. Again, unfortunately, when it did so, the results of string theory's predictions were off by 100 orders of magnitude. But the theory's champions always had an easy out—just change the value of one or another of those dimensions, and you could force any result to fit.

As the years of our new century advanced, it became increasingly obvious that the string business was leading nowhere. String theorists predicted that there are 10^{100} ways that reality can manifest—a vagueness that made most physicists throw their hands up and consider the string business to be useless.

And that's the current state of affairs in our search for a Grand Unified Theory. It dead-ended before the Second World War, though this fact has taken some decades to sink in.

Naturally, the authors say that these models were doomed to fail because theoreticians always tried to fashion a theory of everything that ignored the section of reality comprising somewhere between 50 and 100 percent of it: the observer. And yet this reality kept tugging at their sleeves. Scientists repeatedly saw powerful links between the physical universe and the life inhabiting it. Experiments went one way when no one was watching, and another when we, the observer, stepped in for a closer look.

Meanwhile, QT showed that space or separation between objects mutated dramatically (or vanished altogether in the case of EPR correlations), and time became suspect as well. Yet no one thought to explain or do anything with all these quirky anomalies. They were instead treated as sort of profound oddities, and got little more than shrugs. Footnotes. Asterisks.

Science wasn't to blame for trying to keep people out of the equation. Humans screw up. Our predilection for error is renowned. Just survey the eyewitnesses to a car or plane crash, and you'll get accounts that reliably diverge. Besides, science works better when you can remove the human factor. Not much benefit can be found in subjectivity. If you want to design a better airliner, do you really want to incorporate peoples' hunches and moods? Quite the opposite; aeronautical engineering requires repeatable tests that lie totally outside the realm of individuals' quirkiness.

But in discarding human personalities and foibles, science also turned its back on the fundamental act of perception itself. What was brushed off as irrelevant was something profound, with roots that predate personality and even taxonomy. In reality, awareness is something deep rather than idiosyncratic. It is basic and permanent rather than transient and dispensable.

Another persistent problem with the Grand Unified Theory attempts was that what works when dealing with parts may not be valid when it comes to the whole. It may seem reasonable to extrapolate from apples to planets to galaxies to the cosmos as an entirety—but there's no reliable basis for assuming that the results will have legitimacy. Here's why.

Consider the elements chlorine and sodium. Chlorine is a poison, and a major component of some of the horrible gases used in the First World War. Sodium is hydro-antagonistic—toss some in a lake and you get an explosion. If you're a water-containing life form handling either element, you're dealing with a pretty brutal item.

Studying their structures, melting points, atomic weights, and all the rest could give no hint of what you'd have if you combined these two elements. But bingo: Let an atom of one bond with an atom of the other, and you get sodium chloride—common table salt. Now, no longer does an explosion rattle the neighborhood when this new compound meets water. Precisely the opposite happens. As part of salt it readily dissolves, leaving the water as transparent and unruffled as ever.

As for that chlorine, now instead of a poison you have a substance vital to life. If all the sodium chloride could be suddenly removed from your body, you'd quickly die. There'd be no way to predict this larger result attained by combining the parts. The whole proves to be unpredictably different, even opposite, from its constituents.

Or consider how our logic system works in everyday macroscopic life. We've devised ways of thinking that work perfectly well when we want to communicate or hunt or build bridges. But because we had no experience with—or need to experience—the sub-microscopic realm, we never evolved mental tools to grasp it. It turned out, the logical processes that work on the everyday macroscopic scale are irrelevant when you get to realities six orders of magnitude smaller.

In everyday life, things work logically because that's what logic was created to deal with. At this moment your kitchen either: (1) has one or more cats in it, (2) contains no cats, or else (3) has partial cats (if they're lounging in doorways and are neither fully in nor out of the room). These are the *only choices* when it comes to the topic of cats and their relationship to your kitchen. No other possibilities exist but these three.

Now consider electrons that are created in one spot and then beamed to another where there is a detector; we'll call this route path A. Along the path lies a series of mirrors that bounce some electrons to make them take a longer route to the detector, which we'll call path B. We'll fire one electron at a time, and attempt to measure which path it took.

We know it must take A or B because if we block both paths, no electrons reach the detector. But when we measure the routes by various methods, a funny thing happens. Carefully labeling the positions, we find that some electrons reach the detector by neither taking path A, nor path B, nor both paths together, nor neither path. Since these are the only choices we can logically entertain, the electrons have done something else—something we cannot imagine. Something that lies totally outside all conceivable possibilities and thus our everyday logic.

This is fact, not speculation. That electrons and everything else in the sub-microscopic world can routinely do impossible things is given a name: They are said to be in a state of *superposition* (which was briefly discussed in chapter 7). This corresponds to existing and acting in all possible ways at once, and even some seemingly impossible ways. It's as if today you went to the bank and also didn't go to the bank and both statements were absolutely true.

Now, if the land of the very small lies outside our logic system, why must the meta-universe, the cosmos as a whole, be any more obliging so far as our thought-systems operate? Rather, we should face up to something that's rarely if ever voiced in modern cosmology: the possibility that the true nature of the universe as a whole has *nothing* to do with the way its parts work, that it indeed lies outside the very characteristics of its components.

That the universe (taken as a whole) *does* lie beyond our logic should be obvious, but somehow escapes the notice of cosmology textbooks. Look at our models: Many say a Big Bang started it all, but have no idea, not the foggiest, how you get an entire universe of matter/energy out of nothingness. The very idea makes no sense whatsoever, even if it may sound okay to the majority of people simply because it's been repeated so often. (The very term *Big Bang* was actually coined pejoratively by Fred Hoyle in 1949 as a way to ridicule the notion as preposterous on its very face.)

But say there was a Big Bang, which much astrophysical evidence supports. Then you must face what existed before that event, and of course this is unanswerable. Even saying the cosmos had a beginning leads immediately to illogic, because then where did *that* begin, whatever-it-was? Isn't it clear that we've set up an unsolvable situation with the echoes of an infinite regression? As Thoreau pointed out, we are like the Hindus, who conceived of the world as resting on the back of an elephant, the elephant on the back of a tortoise, and the tortoise on a serpent, and had nothing to put under the serpent.

Abandoning any natal moment for the cosmos doesn't help. Say everything is eternal (which is probably true, if recent "infinite universe" evidence is any guide). Can you picture that? No one can. The same applies to the cosmos having a boundary versus it being infinite. Neither provides an answer of any kind. Neither fits within the workings of logic or the way science is fashioned.

Isn't it obvious that a situation that always yields no answer, that invariably ends in utter mystery, is being tackled by a process inadequate to the job?

Our current models do not work. They do not answer anything at all. They take some of the parts—the 2.73-degree cosmic microwave background radiation, say—and attempt to fashion the Whole Picture around it. But they aren't successful. No result produces satisfaction.

Biocentrism greatly improves the situation, clarifying our understanding of what's going on by bringing life, the observer, into the picture. And because absolutely everything studied, perceived, observed, thought, or conjectured occurs in the matrix of consciousness, the latter *must* be part of the Big Picture, perforce. Is anything more obvious?

Doing so, we find that the squirrely nature of space and time suddenly makes sense, because they are tools of our mind, a way to frame and order what we experience. They are the language of consciousness. It's our way of navigating from point A to point B, of keeping appointments and all the rest. We ourselves carry around time and space like turtles with shells. But seeing how it works, at last, yanks the rug out from under the unsatisfactory conventional models of reality. It explains Heisenberg's otherwise bewildering truth that one cannot measure both momentum and position. It explains why science has found that space is relative to the observer, as is time. Bring life into the picture, and we know why these things work the way they do—it finally all makes sense.

We have to confess, as if at an AA meeting, that we've long been trapped by the habit of visualizing everything in a space and time framework. We couldn't help it, and it was fine so long as we wanted to measure the length of a bridge or the distance to the Sun. But when we wanted to understand the cosmos as a whole, and our lives, and our place in the universe, we see now that we were using a scaffolding that warped and wobbled like smoke in a dream.

We tried to visualize the universe as a giant ball hovering in space. But a ball located where? And what was outside it? And we maybe saw it spatially extending infinitely far in the distance, except we couldn't picture that at all. And we placed it in time as having started long ago while knowing it couldn't, because it had to be beginningless, and we couldn't picture that, either. So our time and space gridwork never actually worked. But we used it anyway because everyone else did. And cosmologists seemed to. And weren't they smarter than us?

So now let's instead crumple up that paper and start afresh, this time with honesty. We must abandon the space and time thing; we know that now. It's okay to picture parts that way—to say that Alpha Centauri is four light-years away. But we can't apply that way of thinking to the cosmos as a whole, any more than we can explain how an electron can take neither path A, nor B, nor both, nor neither.

Instead, we ponder the Whole of Existence, or *Being* as Parmenides called it, and realize that life, consciousness, awareness, and perception are front and center, playing a central role in the experience. We watch the quantum experiments and realize that the physical world is deeply linked with our awareness.

So far, so good. We study books about the brain, and realize that all we see, feel, touch, and hear is occurring strictly within the mind. Here we stop and catch our breath. The universe we perceive is inside our mind. True, the brain exists within the universe, and it is nourished by the Sun's warmth. Yet that Sun has no brightness or warmth outside of our perceptions. (On its own, if there's such a thing, it's invisible, and emits merely electrical and magnetic fields, but no warmth or brightness.)

We sit down and try to let this in. What the cosmos is, is a correlative amalgam of nature and the me, the observer. We are a single essence. We are transactional. We see now why sages have been talking about "The One" since at least 2400 B.C.E. They saw it, they got it. To Zeno, too, it was so obvious he pulled his hair out trying to make everyone grasp that a single event is unfolding—this Oneness acting with limitless energy and animation, effortlessly, never running out of steam.

Renowned physicist Erwin Schrödinger, in his 1944 book, *What is Life?*, wrote that "consciousness is never experienced in the plural, only in the singular."

When discussing the prevailing Western belief in multiple souls, he wrote, "The only possible alternative is to simply keep to the immediate experience that consciousness is a singular of which the plural is unknown."

Indeed, he believed that "the *is* [exists as] only one thing and what seems to be plurality is merely a series of different aspects of this one thing, produced by a deception (the Indian *maya*); the same illusion is produced in a gallery of mirrors, and in the same way Gaurisankar and Mount Everest turned out to be the same peak from different valleys."

It would be well, perhaps, if we were to remember the words of Omar Khayyám, who "never called the One two," and of the old Hindu poem, "Know in thyself and All one self-same soul; banish the dream that sunders part from whole."

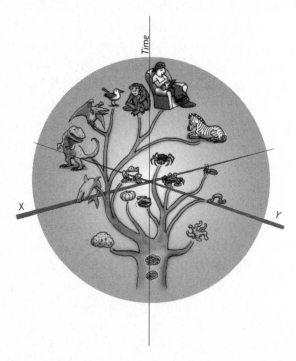

Space and time are not the hard, cold walls we think. Our individual separateness is an illusion. Ultimately, we are all melted together, parts of a single entity that transcends space and time.

We get part of it now, biocentrically, but realize that we'll need some kind of new way of perceiving if we ever hope to grasp it all. Having let go of the time and space framework, we have no new language that can let us be intellectually comfortable with the oneness. That's because when we revert to symbolic language, paradoxes *must* arise.

We tend to forget that all knowledge is relational. Up makes no sense if there isn't also a down. Easy can't exist without the accompanying concept of difficult. The information stream is truly analogous to its components in being codependent, like the ones and zeroes of all digital data. There's on or off, yes or no, each needing the other to have any meaning or usefulness. Utilizing these simple correlated opposites, our minds understand the world.

If mind and nature are correlative, where do separate minds meet? Is it all one there, too? If we're totally relaxed, could we watch other people, and their actions and ours, and feel the same effortless power animating us all?

Can we *really* relax and see our own daily activities as unfolding all by themselves, and that the truer picture is not that of just "little old me" in a vast, scary universe? Can we let that go, along with our fears of mortality?

And if nature/consciousness "oneness" takes getting used to, if you find it strange or hard to accept, we might simply remember that the old, classical, standard cosmological model with its beginnings and infinities and contradictions made far less sense. So in being offered a newer life-based paradigm and being asked to discard the previous one, it's not as though we must now choose illogic over logic. The old model was always illogical. It stayed alive mostly through inertia. It had the advantage of familiarity.

Biocentrism cannot offer an ultimate answer to all cosmic mysteries. But at the least, a life- and consciousness-based reasoning *must* be closer to reality simply because it doesn't ignore existence's most fundamental aspect.

If you can grasp and maybe even *feel* the truth that the mind's algorithms create all we experience, you'll know that the same power that makes our hearts beat also animates the world.

If so, we've found our Grand Unified Theory.

YOU'RE DEAD. NOW WHAT?

17

Because I could not stop for Death—
He kindly stopped for me—
The Carriage held but just Ourselves—
And Immortality.
—Emily Dickinson, "Because I could not stop for Death";
Poems, Series One (1890)

Here is where we tell you what happens after you're dead. Seriously.

Okay, it's not so serious, because you won't actually die. Probably 72 percent of readers will, at this point, assume this chapter will be a load of cow fertilizer, because who can say for sure? Well, hang in there and decide for yourself.

Before we get into the biocentric explanation, we need to take a little side trip. First we'll briefly rewind to the standard everyday view of mortality, which isn't pretty. At the least, it's an awkward conversation stopper, whose only good feature is its tendency toward brevity. Essentially you drop dead and that's the end of everything. This is the view favored by intellectuals, who pride themselves on being stoic and realistic enough to avoid taking cowardly refuge in Karl Marx' spiritual "opium"—the belief in an afterlife. This modern view is not a cheerful one. When Woody Allen was asked his opinion of death, he said, "I'm strongly against it."

If instead you are anachronistic enough to be Sunday school religious, your scenario will find your soul journeying to heaven or hell, where you will remain forever; or else purgatory, which is kind of like a dentist's waiting room. If Eastern religion is your thing, you assume you'll instead wake up in a baby's body, destined in a few short years to once again memorize the multiplication table.

On a science level, a dead body may be fascinating, but to most it's morbidly unappealing. Few medical students or funeral home visitors pause to philosophize: What exactly is this mass of protoplasm? Our visual sense was designed to perceive particular wavelengths of electromagnetic energy—in this case the lifeless body's gray color—and science can reveal the deceased's weight, but physics says that this motionless body in front of us is really a vibrant flux of energy, of electrical fields and mass/energy equivalency. If it could be fully utilized, the corpse's energy would keep all the lightbulbs in the United States ablaze for two and a half years. (This would necessitate having the dead body contact its antimatter analog, so that your late friend George would have to be laid to rest together with an antiGeorge.)

It's also true that almost nothing there is solid. More than 99.9 percent of the matter in a body is confined to its infinitesimally small atomic nuclei, which if combined would be a speck too tiny to see. Is that invisible, trifling dot truly all that exists when this person's lifetime of hopes and dreams are clinically appraised? Science seemingly makes death a bit trivial, even anticlimactic. There's got to be more going on than meets the eye.

All right, then, what is it that meets the *intuition*? We may not think about it very much, but logic and perception are two very different animals. Sometimes they coincide. For example, we all enjoy hanging around a bonfire. On a logical level this makes sense because flames are colorful, animated, ever-changing, and very much enchanting. Thus, fire is logically appealing as well as attractive on a gut level.

But now consider a trip to the countryside on a moonless night. Strictly on a logical level, nothing is particularly beautiful. It's visually a matter of white points on a dark gray backdrop. The only other visible entity is the Milky Way, a mottled grayish-white band crossing the heavens. Why should this be visually special? And yet everyone who steps out of their tent on an overnight wilderness camping trip experiences a wordless rapture. It's an intuitive feeling, not at all a logical thing.

The same is true of a total solar eclipse. Everyone has seen photographs of the black cameo of the Moon blocking the distant Sun behind it. The photos look interesting to the same degree that watching beavers build a dam is interesting. Yet the actual in-person experience of totality makes folks weep. Some utter animal-like exclamations. It's life changing. A partial eclipse, which requires eye protection, does not accomplish this. Happening only every 360 years on average for any given place on Earth, a solar totality is rare, but that's not the explanation, either. Something about the Sun and Moon and Earth forming a straight line in space creates a feeling—what hippies would call a "vibe"—that has no logical correlate, though it almost knocks onlookers backward.

The point to all this is that we humans perceive and evaluate the world using a variety of tools. The tool of logic is sometimes the most appropriate in a given situation, while at other times the superior instrument is direct perception. Both processes happen spontaneously. Sometimes they agree with each other. We're introduced to someone new; their good reputation has preceded them and logically we expect to like them. Upon shaking hands and making eye contact everything feels warm and comfortable on an intuitive level and it all meshes. We immediately decide that this person is okay. But once in a while a meeting is awkward. The person's vibes, to revert once again to hippy-speak, feel weird or judgmental or angry or unpleasant in some way. And yet "on paper" we expected them to be peachy. Our intuition is in conflict with our logical minds. In such cases, which should we trust?

In practice, most of us trust our intuition above everything else. Now, the only reason we're investing so much time on this issue is to illustrate a truth that may sound unscientific but is actually inarguable: Intuitiveness is real, and usually reliable.

If the reader may indulge a bit more of this excursion away from the purely scientific, this *instinctive-level perceiving process* reaches an exquisite perfection when an observer is no longer blurring between the various levels of logic and instinct, but focuses exclusively on one tool or the other. Solving a difficult math problem requires a one-pointed logical focus and total lack of distraction, where any intruding emotional feelings or digressions to one's surroundings—the sunset outside the window, say—is a hindrance.

It works the other way, too. The sage, the mystic, the enlightened person is thoroughly freed from the logical mind and clearly sees nature with full intuitive focus. He is one with the environment. In that state, people are not perceived as "others." They are seen as the single oneness. In each stranger's eyes, the sage sees "the face of God." The sage also directly perceives that this single amalgam of unity, the "Being" described by some ancient Greeks like Parmenides, is deathless. If all is an eternal existence of life and nature—the true "self"—what can die? Birth and death are apprehended as illusions, and this perception is accompanied by conviction, a sense of certainty. It is cognized as a *recognition* of reality rather than as an acquisition of a new idea.

One little, additional item before we leave this direct-perception business. After all, because biocentrism avers that

YOU'RE DEAD. NOW WHAT?

nature and the observer are correlative, direct perception is inherently a valid process, as every observer is already plugged into the essence of the cosmos *and does not stand apart from it.* He or she feels its truths on some deep level. How could it be otherwise?

So our final intuitive exploration asks this question: What is a dead body like?

Those who have stood next to a corpse, perhaps of someone dearly beloved, know that it feels very different from the living person. Even when someone is asleep in the backseat during a long drive, you can sense their presence. Everyone has a unique "feeling" to them. This isn't New Age mush, even if it sounds a bit like it. It's just that we're so accustomed to the intangible aura of our friends and families, we don't usually attend to this aspect of the people we know. But when alongside their lifeless body, it's strikingly obvious that Mom or Bill is simply not there. The feeling is so dramatically different it's disconcerting, even creepy. It's not simply that they're no longer moving or breathing. Indeed, funeral directors will tell you that people routinely say, "Mom is gone, that's not her anymore."

Some vibrant, conscious quality that was the actual person we knew and loved is now absent. In short, *people are not their bodies.* What we fail to do is apply this revelation to ourselves. If they are not their body, then neither are we.

So self–body identification is the first mistake we make when we're trying to probe that nagging issue of mortality. We look at our limbs and say, "That's my hand," but who is the "me" who possesses the hand? We could theoretically chop off all parts until we're only a brain in a bottle somehow kept alive with nutrients, yet we'd still feel, "Here I am. I'm still fully me!" And if that "me" feeling fully endures regardless of how much of your body is gone (ask any unfortunate multiple-amputee war veteran), what if the electrical swarm that constitutes consciousness could be maintained in some kind of futuristic plasma container? Would we not then totally realize that we really are not our bodies?

Animals have no trouble with this. Your cat has no idea what she looks like. She doesn't even know she's a cat. She doesn't imagine she has a body of any sort. She'll clean herself not because she's body conscious but because that action comes naturally and instinctively; it feels like the thing to do. She may lick you, too, if you get your hand in the vicinity.

The body dies. The real "me" does not. Or at least, once you've clearly seen that you're not your body, the issue of what happens to the "me" becomes an entirely separate matter.

Let's bring biocentrism back in. The feeling of "me," of consciousness itself, could be considered a 23-watt energy cloud, which is the brain's energy consumption in producing our sense of "being" and its myriad sensory manifestations. Energy, as we learned in high school physics, is never lost. It can change form but it never dissipates or disappears. So what happens when those brain cells die?

First off, never forget that the mind's algorithms create your idea of your brain, while specific sensory architecture create a brain's *appearance* when you dissect one in medical school. We've already fully seen that neither space nor time are real in any sense except as appearances or tools of the mind. Thus, anything that seems to occupy space (like the brain or body) or endures in time (again, the brain and body) has no absolute reality, but only an apparent one created by the mind. Let the mind change its neuro-chemical meanderings, and the space and time appearances vanish like dissipating smoke.

In biocentrism, it's clear that correlative spheres of reality don't have absolute spatio-temporal order independent of the observer. The observer alone creates space and time, so from the get-go we cannot pretend there is some absolute space-time matrix in which a body dies. Indeed, in the absolute sense, we can't even say (absent the observer) what events have occurred before others. Time and sequencing are meaningless to nature.

Because space and time are tools (concepts) of our mind, not real external objects like cucumbers, all our knowledge is relational and is based on these spatio-temporal relationships. We cannot comprehend *anything* outside this spatio-temporal system of thought. Nature (or the mind's) pre-thought structure, which for lack of a better word we'll simply call *information*, has no spatio-temporal meaning before the algorithms of our mind impart order. Thus they cannot be thought of as "going away"—which requires the temporal concepts of before and after.

In short, the very idea of death, or becoming nothing, is empty of meaning. Becoming nothing may seem like a tangible concept, but it is actually as meaningless as the word "it" in the phrase "it's a nice day." It appears linguistically, but not in the actual physical universe. The information that constitutes our selves or conscious awareness exists outside our linear spatio-temporal thinking.

Because time doesn't exist, there is no "after death" except the death of your physical body in someone else's now. Everything is just nows. And because there's no absolute self-existing space-time matrix for your energy to dissipate, it's simply impossible to "go" anywhere. You will always be alive.

We experience only a reality where the algorithm creates the burgeoning sense of self or nature, the way a needle on a phonograph record manifests a sound. The process turns this information into the three-dimensional reality we know and experience, such as the music being played at any given time. All the other information on the record (nature or cosmos) exists in superposition, as potential.

Any causal history leading up to the "now" being experienced can be thought of as the "past" (i.e., the songs that played before wherever the needle is), and any causal events that follow the "now" (i.e., the "present") occur in the "future" (i.e., the songs/music that plays after wherever the needle currently is), but really, only the *now* exists. The other seeming states of past or future materialize only when the mind has created its 3D reality. The before-death state, including your current life with its memories, goes back into superposition, into the part of the record that represents just information.

In short, death does not actually exist. If we wish to consider the nature of any change that is unfolding upon bodily dissolution, it could be thought of as a reboot. Certainly a positive experience, a freshening up. At death, we finally reach the imagined border of ourselves, the wooded boundary where, in the words of the old fairy tale, the fox and the hare say good night to each other. But if time is an illusion, so too is the continuity in the connection of nows. Where, then, do we find ourselves? On rungs that can be intercalated anywhere, "like those," as Emerson put it, "that Hermes won with dice of the moon, that Osiris might be born"?

Once again, past and future are ideas relative to each individual observer. You know you had a grandmother who also had a grandmother. These are ideas, true, but it's not a stretch to assume they each had their own bubble of spatio-temporal reality—just as you assume that the people around you each experience spheres of time and space realities, even if all are fundamentally one with nature and—in the deepest sense— indistinct from the whole.

There is no ticking matrix of "time" between these spheres because there's no such thing as time except as a concept in the mind of each individual observer. What's most important to remember is that past, present, and future *between* these separate bubbles of reality have no meaning. So neither does any kind of death followed in time by rebirth.

Many believe in reincarnation, and on a limited level it may not be wrong. But in a truer sense, what is it that can be reincarnated when there's no death to begin with? Not to mention no actual, separate individuals in the nature/consciousness correlate that constitutes the single eternal Being. The bottom line, the takeaway conclusion for anyone who fears death, is that your consciousness is never discontinuous.

No wonder Parmenides, 2,400 years ago, figuratively ran down the streets of Elea trying to spread the happy news that reality is actually simple and safe. Along with Zeno, who lived down the block, he was flummoxed by the notion of a multifarious cosmos rife with mortality, which was starting to gain favor among the newer Greek philosophers who were showing signs of overthinking everything.

Like Parmenides, you and I know that time doesn't exist except as ideas in the now. Thus, "past" and "future" are illusions. And with it, any time-dependent notions, including the biggest and baddest: that you who exist as awareness, will ever cease to be.

As Einstein wrote, shortly before his death in 1955, "For we convinced physicists, the distinction between past, present, and future is only a stubbornly persistent illusion."

GRAND ILLUSIONS 18

We have the need to fool ourselves continuously by the
spontaneous creation of a reality . . . which reveals itself to be
vain and illusory.
—Luigi Pirandello, autobiographical sketch in *Le Lettere* (1924)

Biocentrism says everything is relational, and it's true.
Most of us labor under the stubborn illusion that
there's a self-existing "rough" without an accompanying
"smooth"—and imagine an actual "out there" that exists apart
from our conscious selves. And that there is an insensate external world that comprises the vast bulk of reality, along with a
separate "little old me" that confronts it.

This illusion makes us feel that the overwhelming majority of the cosmos is inert and lifeless. That our separate spark
of personal experience flares briefly as a parenthetical bit of
animation surrounded by a permanent dead void. No wonder it feels so freeing when the cosmos is correctly seen as
a life-centered entity. Then its attributes are perceived as

wondrous, and the "I" feeling associated with being separate from others fades away.

Acquiring this mindset might start with grasping that there's much more to life and existence than we previously assumed. "Name the colors, blind the eye," goes an old Chinese proverb. If you only "see" the hues you've assigned labels to, you're missing the entire range of color sensations. Logic alone creates these illusions and they extend everywhere. For example, we believe our brain rules the body. But we might just as easily imagine our stomachs and livers—craving glucose and energy—having "grown" a brain to subserviently hunt and connive and find food to service these organs. In reality, nothing is separate, nothing rules anything; it all goes together. This escapes us mostly because of the intellect's machinery, which creates symbols for things after dividing up reality into a relatively small number of disparate parts. At the linguistic level, separate objects exist only if we have a name for them. This in turn makes us miss most of life, and very definitely comprises a more confined experience than perceiving oneness.

We may not think much about them, but illusions are commonplace companions. If we look at some of them and see their prevalence, we may feel more open to jettisoning habitual views. They often start with routine sentence constructions that involve "I" pondering reality or "I" observing galaxies. So when probing illusions, we shouldn't ignore the most intimate conundrum—ourselves. The "me" feeling stems from a cloud of electrical consumption in our head that has sent brain researchers and philosophers into a more or less permanent tizzy. Is the "I" who daydreams and orders a drink an entity separate from the "I" that performs various functions within the kidneys? Where does "I" begin and end?

If we try to be strictly objective and scientific, we might define "I"—the self—as the body, and say that the epidermis comprises its boundaries. In short, "I" am everything within that waterproof bag of skin. If we instead use each person's subjective sense of "me," then definitions get much trickier. After all, you probably never felt your toes to be fully "myself" because you'd always refer to them as "my toes" as if they were possessions.

My arms, my legs, my liver. But who is the possessor of all these—the "Me"?

This is the central question posed seventy years ago by the great South Indian sage Ramana Maharshi. He tirelessly championed the "Who Am I?" method for unlocking the deepest secrets. It's simple, he'd insist. Don't trouble yourself with endless questions about God, existence, destiny, and all the rest. Instead, find out who is the person who wants to know such things. Who is the one experiencing these mental torments?

So, as a meditation, a person was encouraged to simply see where the "I" experience arose, and what it was. *Who am I?* Well, the "Eureka!" moment and all its benefits would accrue only to whomever went through the effort of "looking within" and seeing where it took them.

A person who made such self-inquiries with all sincerity and good effort ultimately could find no one home. He or she would discover that there is no separate individual self, only a stream of thoughts. Or, put another way, upon this realization, one would clearly see that the "self" was either nothing at all—or the entire cosmos. Thus one of the all-time biggest illusions is the existence of a separate mortal "Jessica" or "Michael"—you, as a stand-alone entity existing apart from the cosmos.

This is why the Maharshi often alluded to the universe as "the Self" with a capital S. As opposed to the false sense of self with a small s, which one imagines as a kind of talking parrot in one's head.

Nor was this perception confined to the Eastern Hemisphere. Eleven years after he won the Nobel Prize in Physics, Erwin Schrödinger wrote:

What is this I? If you analyze it closely you will, I think, find it is just a little more than a collection of single data (experiences and memories), namely the canvas upon which they are collected . . .

Yet if a skilled hypnotist succeeded in blotting out entirely your earlier reminiscences, you would not find that he killed you. In no case is there a loss of personal experience to deplore. Nor will there ever be.

Our point, again, is that our biocentric conclusions that there is no death, no time, no space, and instead a single living entity, which precludes a stand-apart dead universe abiding separately from life and consciousness, is a science-based reality, but it's also the conclusion that anyone would arrive at on their own if they merely thought things through, or quietly contemplated what was going on inside their minds.

And what's going on encompasses both the interior of one's body as well as the exterior universe. In reality there is no distinction. We are an amalgam, an entity consisting of the outside world and the body/mind. Like trees whose roots branch down and outward and whose topmost, thinnest branches reach up and outward, we too are it all. Air, water, electrical current, the planet itself, and our body/minds, all built as an interrelated living organism. We didn't arise from the universe. We don't even merely *express* the cosmos. We *are* it. Its air and water is our being and we do not live without its inseparability with us.

In stark contrast to the "little old me" feeling of the "I," illusions can also be physically enormous. In 2012, a team at the University of California, Berkeley studied 900,000 galaxies and found that large-scale space shows no sign of warping. The conclusion? This flat large-scale topography indicates that the universe is probably infinite, since a finite cosmos would display a curvature in its space-time, caused by the enormous mass of its combined galaxies and dark matter. This new discovery indicates that the cosmic inventory of galaxies and planets is endless. In April 2013, Debra Elmegreen, then president of the American Astronomical Society, shrugged it off to one of the authors who'd asked what she made of this news that the visible cosmos is enveloped in an infinitely larger matrix: "Even if we can only observe a very small fraction of the universe, that's plenty to keep us busy."

But she slightly misspoke. It's not a very small percentage that's observable. You see, any fraction of infinity is zero. It means we cannot see even a few paintbrush strokes of the celestial masterwork. Thus, as briefly noted in chapter 1, all we can ever hope to study is *zero percent*. And when a sample size is zero, no conclusions are trustworthy. Thus this illusion extends to everything we think we know about the cosmos.

Take the idea—popular among some cosmologists—that everything started from nothingness. In short, the positive attractive force of all mass and gravity is balanced by the negative repulsiveness of dark energy. The plus and minus cancel out. Thus, some theorists conclude with straight faces that the universe is fundamentally nothingness.

Is this helpful? Is it valid reasoning or technobabble? Can you really get something from true nothingness? Calling things positive and negative and then saying they cancel into blankness doesn't mean they are *actually* positive or negative except as mental classifications.

In truth—and the reason we're even "going there" when it comes to the largest-scale studies of the cosmos—is to make clear that the Great Everything continues to hold ineffably profound mysteries. Werner Heisenberg once said, "Contemporary science, today more than at any previous time, has been forced by nature herself to pose again the question of the possibility of comprehending reality by mental processes."

No physicist can escape the elusive issues of how the universe materialized or if it even had a birth. The Big Bang's zero moment remains an utter enigma regardless of whether one holds to the classical model of a cosmos separate from our awareness, or not. Even calling the Big Bang a "beginning" to the cosmos doesn't carry us beyond square one because no one knows anything about the possibly infinite entity from which it arose. We can only guess about the larger cosmos, whether it be the classical realm beyond the observable horizon of light-speed expansion, or the biocentric realm beyond the mind's algorithms.

Bedrock issues—Was the universe born? What's its size? What's it really made of?—remain enigmas even in today's standard non-biocentric models. The inscrutabilities of consciousness, shared by classical physics, quantum mechanics, and biocentrism, add an additional mystery ingredient to the stewpot.

Even if you back up to the simplest concepts of something and nothingness, one of these had to be created, whether at the Big Bang, by God, or whatever your personal genesis inclination happens to be. If you assume a real, independent world out there, then creation is exactly the problem you run into with the Big Bang, barring some theoretical hand waving. If there was nothing before it, then everything was just suddenly created from nothing. It should be obvious by now that contemporary concepts about the true nature of this universe are sleepwalks in a dreamland of illusion. How to escape them?

We start by seeing that there simply is no real world "out there" beyond us. The Big Bang is part of a relational concept, a logic related to the observer. Nothing physical has to be created because everything is the same always-present mind-concept.

According to biocentrism, we cannot fathom *anything* outside the spatio-temporal logic comprising the mind's algorithms. If you ask how *that* came to be, whether it had a birth, or all such imponderables, you're back to the *Being* of Parmenides and the mystics. If it can be apprehended directly in "enlightened" states, so be it. But in the science and logic business, if we want to know what to make of such unexplorable meadows, there's no reason it must be either titillating or depressing. The ancient

Greeks weren't bothered at all; they generally found futility amusing. If something lay clearly beyond their ken, they'd laugh and pour another glass of wine.

The first illusion-busting step involves tossing out today's dead universe paradigm, which is as antiquated as all those turtles that once supposedly supported the flat Earth beneath our feet. Even then we may still be teased by things ungraspable via the matrix of logical symbolism, the medium of language.

The second illusion-busting step is to toss out the "I" feeling. Even if we grasp that this "me," ourselves, is what creates the framework of time and space, that's probably not enough to impart the full, exuberant experience of unity. Indeed, as our powerful technological instruments extend both our perceptions and our illusions (as when we use our telescopes), we're filled with awe at the universe's vastness but also invariably feel even smaller and less consequential. Thus, modern science knowledge rarely helps us get closer to what's really going on when it comes to the Whole Picture.

It's a vast realm, that of visual phantoms and optical illusions. Ours is a funny old universe, with real things that cannot be seen, like love and neutrinos and dark matter. Yet, conversely, it also has dramatic-looking entities that lack any physical existence at all. We've awakened in a hall of mirrors.

Enchanting indeed are their countless reflections. Still, by penetrating the veil of appearances, biocentrism opens an entire new world of possibilities.

WHERE NEXT?

19

My striving eye Dazzles at it, as at eternity.
—Henry Vaughan, "Childhood" (1655)

I n a book devoting many pages to disproving time, this chapter's title might seem bewildering. But as we now wrap up what we take away from all this, it's not that we should "uplevel" in everyday life by speaking or behaving as if we have the luxury of living alone on an island. We have appointments to keep. We live in a society based on a shared notion of time and have to act accordingly if we're not to be locked away in a psychiatric ward.

A goodly percentage of us seek "a meaning to life" or the search for what's real, if only because we want to be fully candid with ourselves intellectually. The biocentric paradigm goes a long way toward aiding our understanding of the cosmos, and its supporting science is already formidable. But if the underlying unity of nature and the observer with all its implications—chief among them the unreality of death—is to gain wider acceptance, it will surely arrive through ongoing scientific verification.

After all, science in the past century has already utterly altered our perceptions of the size of the cosmos (before 1928, when Edwin Hubble pinned down a more accurate distance to spiral nebulae, it was largely presumed that only a single galaxy existed), demolished the belief in locality (that physical effects can only be caused by actions of nearby objects or forces), crossed off Mars as a probable planet with life, and unveiled countless other revisions to what used to be mainstream views of reality—especially between 1905 and 1935, when Einstein and then quantum mechanics changed physics forever. Some of these revelations subtly but decisively altered everyday life.

Not all of the new knowledge had a positive psychological outcome on the mainstream mindset. Science's ever-growing twentieth-century assumption of a dumb, random universe, in which life arose by chance, had the secondary effect of isolating the human psyche from the cosmos. It probably made almost everyone feel inconsequential and lucky even to be alive. This, together with the growing abandonment of religion, probably led to a sense that, in a cosmos ruled by accidents rather than by plan and/or perfection, we humans need to exploit the environment and grab what we can. A universe fundamentally separate from ourselves, in which we arose by some fluke, is also a cosmos that could turn on us at any moment. It has set up an antagonistic outlook: humanity against nature.

Our current models can't help but make us feel isolated from the cosmos, and vulnerable, with ongoing effects on our routine outlooks. Thus, twentieth-century cosmology has proven more than merely incapable of providing any picture of reality that makes sense. It has also fundamentally alienated us from nature. Therefore, yes, science can and does influence us experientially and emotionally, not just intellectually.

This is why we hope and expect that further research will not just support the biocentric model of the cosmos, but ultimately be incorporated into our worldviews. On a personal level, for each of us, what benefit might *specifically* accrue?

First, of course, truly seeing the reality that we are one with nature and not apart from it, that consciousness is correlative with the cosmos, immediately helps ameliorate our war with the environment. You cannot wage war on yourself. (Well, maybe you can in some way, but you get the point.) Certainly a kind of peace or satisfaction must arise when one sees that our very selves are intimately interrelated with the galaxies. At the least, it must create some form of relaxation as opposed to an ongoing psychological conflict with our surroundings.

Second, there is logical satisfaction in having a worldview that finally makes sense. All the nagging quantum experiments that point to the importance of the observer, all the bothersome reasons a time- and space-based picture of reality doesn't hold water—it will be good to jettison all the conflicting science oddities most people shrug off as due to their not being smart enough to understand physics. We want our science to work, even on the largest-scale issues. Now it can.

Third, this view suggests tantalizing new directions for research, a combining of biology and physics that is long overdue. The immediate thrusts are already taking place, among them the quest to see how quantum mechanics applies to the macroscopic world, our everyday realities. Because QT already amply demonstrates the intimate link between observer and observed, and shows the connectedness of objects seemingly separated by any amount of space, it will be fun to watch the QT consequences on visible objects rather than sub-microscopic entities.

This is already happening. First, though, one should know why there are good reasons QT's effects are dramatically "in your face" when it comes to the behavior of small items, but far less obvious when encountering vast collections of atoms, and not yet possible at all when we observe the Moon or a locomotive. It has to do with the wave nature of everything, because all matter is fundamentally composed of waves, or at least behave that way when we do the right experiments. The waves of light, electrons, and other small objects are very small and coherent, meaning that when they're observed they exhibit properties like polarization and frequency. The weird-seeming quantum effects, like entanglement and tunneling (objects instantaneously passing through classically impenetrable barriers and materializing on the other side), show up routinely with objects of this size.

Actually, quantum effects *do* visually appear in our everyday macroscopic world. The swirling colors on a soap bubble and the lovely hues of peacock feathers and seashells are examples of diffraction, a quantum wave effect. Even larger (longer-spaced) waves can show quantum effects, as when radio station signals bend around solid objects to be heard in places where classical physics would have deemed impossible.

But when we look at a boulder, we now have an enormous collection of disparate waves, because so many separate atoms comprise the object. Quantum effects still happen, but their probabilistic natures make it a long shot that they all will have their wave-functions collapse in the identical way, especially if it's an unlikely way. The next time you walk into the kitchen, the fridge may have vanished because it's rematerialized in the White House. It *could* happen. Science shows that it's not impossible. But the likelihood is that it wouldn't occur until so many other, much more likely, aggregate wave-function probabilistic collapses had come and gone, that it wouldn't be witnessed until long after the expected human tenure on this planet.

What's important is to remember that tiny objects have easily detectable waves that can interfere, cancel out, or amplify—unlike, say, baseballs. You can't have a baseball hit another and reasonably expect to see both disappear. So the aggregate of all waves makes detecting quantum effects on the everyday visual world much more challenging. In the quantum realm, objects simultaneously exist and don't exist, or can be visualized as behaving in various mutually exclusive ways simultaneously. But in the classical everyday world, it's an either/or situation. Unlike an electron, a cantaloupe is either here or there but not in both places at the same time.

But why? At what size, or under what conditions, do we get a transition from quantum to classical behavior? Many think "decoherence"—the loss of quantum weirdness and both-states-at-once reality—is quickly brought about by an interaction with the environment, so that the larger the object, the faster this happens, as so many atoms are involved. Others think it may be gravity itself that causes the switch to the classical world. Still others posit that some quantum states like momentum are more resistant to loss of coherence than others, and that there's a kind of Darwinism at work where the most stubborn properties retain quantum-ness more persistently. But others think that quantum effects can keep being seen in ever-larger objects, on the visible level. All this is relevant to biocentrism because it reveals the tangible indispensability of the observer with the so-called external world, as well as the nonreality of space and time.

The quantum and classical worlds seem very different, but how physical objects change their behavior to switch between them is currently still unclear and the subject of intense modern investigation. Recently, physicists have hit upon new alternative explanations that once again bring things back to the observer's role. This idea is that any physical system displays quantum behavior when observed with very precise measurements, but will mutate into a classical system as soon as the measurements get too coarse or fuzzy—which they do when dealing with the aggregate of so many particles or photons as found in the visual world. In other words, it's the coarsening of measurements that forces the so-called quantum-to-classical transition. If this is true, then observer dependency does indeed call the shots on all levels.

In the second decade of the twenty-first century, the main research problem in scientifically getting to the bottom of all this was finding that coarse measurement did *not* reliably produce the change to classical behavior, leaving researchers unsure of the exact required parameters needed to definitively bring about the quantum-to-classical transition. However, in a 2014 study published in *Physical Review Letters*, physicists discovered that "measurement" is actually a dual process. Contrary to previous assumptions, the final detection is not the only component. Rather, the full gaining of information also entails setting and controlling measurement references such as timing or angle, which are vital for our minds to really grasp what's happening. The physicists found that when these are controlled, the transition to our classical physical world is invariable and inevitable. It all keeps the observer's knowledge firmly in the scene, in how the cosmos manifests itself.

Thus, we're still learning when it's valid and appropriate to jettison our long-held logic-based thinking, which in turn is rooted in local realism. To review from way back in chapter 7, this classical view asserts that an object is "there" regardless of our measurement of it, and that the object always carries with it all the information needed to determine how it behaves. If our detector shows that it exhibits a particular behavior, classical thinking would have us imagine that we could figure out how it acted before we observed it—for example, which path it took to get to where we now see it.

But biocentrism and its nonclassical thinking works much better, by showing that information that's not specifically obtained by the experimental apparatus does not in fact have any real independent existence or self-existing history. Thus, the particle actually didn't have any sort of "path" before we observed it. Nor does it even have a path when we do observe it—unless our experiment is designed to look for a path. The key is that the reality of that particle *is* our observation.

These days, quantum phenomena like entanglement or tunneling are of huge interest to researchers, not because many care about proving the nature/mind correlation of biocentrism, but simply to exploit these properties for commercial purposes. Chief among them is the potential for much faster responses in the next generations of computers, since QT phenomena operate in the no-time realm of instantaneousness, rather than being limited to lightspeed, as electricity is. Thus, count on QT's realities being increasingly exploited for practical purposes in the years to come.

Imagine how juicy all this becomes as we start to observe these observer-dependent effects in the everyday, macroscopic world. The search for such freaky quantum effects on the visible level is actively under way, with success now reported several times annually in various laboratories around the world. For example, in a 2010 *Nature* article, a team of University of California physicists demonstrated quantum effects on a visible mechanical system, a kind of tiny drum with movable parts just barely visible to the eye. The main problem had been finding a way to sufficiently cool all the objects' atoms to their "quantum ground state" near absolute zero. Once that temperature was achieved, the researchers created a superposition state of the "drum's skin" where they simultaneously had an excitation in the resonator and no excitation in the resonator at the same time. The drum beat, and didn't beat, simultaneously.

Recently, potassium bicarbonate crystals exhibited entanglement ridges half an inch high, demonstrating that quantum behavior could nudge far into the ordinary world of human-scale objects. In 2013, the double-slit experiment was successfully performed with molecules that each comprised 810 atoms. Also that year, a 5,000-atom molecule successfully displayed wave–particle duality. This gargantuan molecule, $C284.H190.F320.N4.S12$, was one-tenth the size of a small virus, showing that these quantum effects are not confined to the realm of the submicroscopic. Each year brings more progress in working with entangled light, particles, and ever-larger assemblies of objects, as science finds the most effective ways of watching QT's magic approach and reach the visible scale.

Among the many people continually working on this is Nicolas Gisin, the Swiss physicist who got the ball rolling in 1997 by conclusively demonstrating the reality of entanglement. Back then he used single photons, bits of light, but in recent years he's used large entangled "flashes" consisting of 500 photons apiece.

Entanglement demonstrates the biocentric principle that neither space nor time exists as independent realities outside of animal perception, and as the entangle-able objects grow larger, the reality of it seems more and more amazing. For example, in 2013, scientists entangled two tiny but visible diamonds. Observing one instantly affected the other. And, turns out, even properties like motion can be entangled, and not just in vibrations like that drum we described earlier.

A few years ago, as reported in the journal *Nature*, researchers worked with two pairs of entangled vibrating particles separated by 240 micrometers. When one pair was forced to change its movement, the other pair did as well. Motion was something never previously entangled. To achieve this feat, the scientists used two pairs of atomic nuclei, so that each had a positive charge and could be made to move through the manipulations of electric fields. Each pair included one beryllium and one magnesium ion, which kept vibrating back and forth toward and away from each other "as if they were connected by an invisible spring." When the researchers changed the motion of one pair by stopping and starting the vibrations, employing fields and precisely aimed lasers, the other immediately and spookily responded in a way that was a perfect mirror image.

Another subject of interest is *what kind* of observation can induce instantaneous changes, like those we illustrated in the double-slit experiment of chapter 7. Would electrons leave their probabilistic state, and have their wave-function collapse, if a *cat* was watching? At present, no one in mainstream science can say they know the answer.

Recently, a team of researchers led by Juan Ignacio Cirac, one of the pioneers of quantum information theory, proposed an experiment to see if viruses can be used in these quantum experiments. Imagine: the entanglement of living beings.

"The most striking feature of quantum mechanics," wrote the scientists,

> is the existence of superposition states, where an object appears to be in different situations at the same time. The existence of such states has been tested [and] . . . current progress in optomechanical systems may soon allow us to create superpositions of even larger objects, like micro-sized mirrors or cantilevers, and thus to test quantum mechanical phenomena at larger scales. . . . Our method is ideally suited for the smallest living organisms, such as viruses, which survive under low vacuum pressures, and optically behave as dielectric objects. This opens up the possibility of testing the quantum nature of living organisms by creating quantum superposition states in very much the same spirit as the original Schrödinger's cat "gedanken" paradigm. [This is] a starting point to experimentally address fundamental questions, such as the role of life and consciousness in quantum mechanics.

It won't be terribly long before observers' effects on physical objects will stop seeming strange to the general public. Perhaps the intimate linkage between the "objective" world and consciousness, observed so routinely at present on the experimental, laboratory level, will someday no longer seem odd even to those taking basic high school science classes. We are almost at that stage now. Yet, what will still need be done is to further explore consciousness itself, research into which, as we've seen, is not even in its infancy. It is possible that a new branch of science and utterly new methodologies will have to be devised, as efforts to date have mainly succeeded only in mapping what parts of the brain control specific areas of awareness.

A further support for the biocentric view is what might be called the global quantum state. As things now stand, objects like electrons are known to have no actual existence, position, or motion until observed, at which instant their wave-function collapses and they materialize in a position or with momentum dictated by the laws of probability.

Now, such collapse requires measurement by a macroscopic device or object, as when we shine light (send photons) onto an object to see what's there. When it comes to a large or macroscopic object, then by definition not all parts of the object are being simultaneously observed. Their properties are thus unknown. Such *incompleteness* is well known to cause decoherence and wave-function collapse as per the results seen in usual quantum mechanics. For example, if we have two electrons in an entangled state, measuring the properties of only one electron without information about the second particle will lead to decoherence, or an apparent breakdown of the entanglement of the two particles. The history we have access to will seem deterministic to us.

On the other hand, if one has information about the states of both entangled particles, experiments show that the entanglement of the two particles is reestablished. Thus, if one could measure the quantum states of all the particles in the universe simultaneously, one would never experience the deterministic world we live in, where everyone is either alive or dead, and where events seem to occur sequentially. Instead, one would directly experience the actual timeless reality, the essence of the overarching cosmos, even if we now visualize it as merely the probabilistic blur of quantum mechanics.

There's more. It is clear that our minds harbor a kind of transcendent consciousness-system in which the normal everyday algorithms can be modulated or even circumvented. Consider dreams, meditation, schizophrenia, or even hallucinogenic drugs. Accessing this hierarchical architecture may allow consciousness to bypass, even if momentarily, the customary spatio-temporal configurations to directly perceive its oneness with the cosmos with which it has always been correlated—but freed of subjectively felt feelings of space or time. "There is," wrote Thoreau, "always the possibility . . . of being *all*." By a conscious effort of the mind, Thoreau made clear that he could stand beside himself, aloof from actions and their consequences; and all things, good and bad, went by him like a torrent. "I may be either the driftwood in the stream, or Indra in the sky looking down on it."

What is not in doubt even in these early research stages is that the observer is correlative with the cosmos. That time does not exist. And perhaps the most cheerful takeaway from biocentrism: Since there's no self-existing space-time matrix in which energy can dissipate, it's impossible for you to "go" anywhere.

In a nutshell, death is illusory. So far as actual direct experience is concerned, you will continue to find what you've always observed: Consciousness and awareness never began, and will never end.

In *2001: A Space Odyssey*, astronauts are sent on a pilgrimage to Jupiter. At the end, Dave Bowman finds himself pulled into a tunnel of colored light beyond space and time to learn the deepest secrets—yet merely finds another riddle. His adventure was an apt metaphor for our long, ancient quest as humans.

As the great anthropologist Loren Eiseley said, "The secret of life has slipped through our fingers and eludes us still . . . so deep is the mind-set of an age . . . [that] the desire to link life to matter [may have] blinded us to the more remarkable characteristics of both."

For over ten thousand years we have looked up to the sky for answers. We've sent spacecraft to Mars and beyond, and continue to build even bigger machines to find the "God particle" or the elusive critical piece of the puzzle that somehow is never solved. We're like Dorothy in *The Wizard of Oz*, who went on a long journey in search of the Wizard, only to return home . . .

. . . and find that the answer was inside her all along.

APPENDIX 1
Brain versus Mind

Exploring consciousness is a far-out experience, especially when it includes an external world that biocentrism shows is actually within the mind. In discussing this, we use the following definitions:

The *brain* is a physical object occupying a specific location. It exists as a spatio-temporal construction, and other objects like tables and chairs are also constructions, which are located outside the brain. (If they were inside the brain, the skull would be awfully crowded, and those chairs would probably damage the brain's delicate neural tissue and interfere with blood flow.)

However, brain, tables, and chairs all exist in the "mind."

The *mind* is what generates the spatio-temporal construction in the first place. Thus, the mind refers to pre–spatio-temporal, and the brain post–spatio-temporal.

You experience your mind's image of your body, including your brain, just as you experience trees and galaxies. Thus those galaxies are no farther away than is the brain or your fingertips.

The mind is everywhere. It is everything you see, hear, and sense—otherwise you couldn't be conscious of it.

The brain is where the brain is, and the tree is where the tree is. But the mind has no location. It is *everywhere* you observe, smell, or hear anything.

APPENDIX 2
Quick-Find Guide

For quick reference, to locate the largest issues explored in this book:

If You're Looking for:	Go to Chapter:
Exploration of time	1–4, 6
Unreality of death	17
Nonreality of space	1, 9, 12
Nature of consciousness	2, 10, 11, 14, 15
Science proofs of biocentrism	4–8, 19
Awareness in machines or plants	14, 15
How knowledge is acquired and transferred	13
Biocentrism as a timeless realization	11, 16
Life arising as a random accident	10, 16
Quantum theory	5–8

INDEX

2001: A Space Odyssey, 9, 140, 199

A
A. M. Turing Award, 141
absolute zero, 84, 195
abstract thinking, 142
Achilles and the Tortoise paradox, 26–28
aether, 77–78
afterlife, 9, 14, 15, 108, 172
aging, 43
Allen, Woody, 172
amen, 77
American Astronomical Society, 185
Amun, 9
An, Liu, 75
antigravity, 4, 80
Antiphon the Sophist, 23
Aristarchus, 12, 13
Aristotle, 13, 26, 34, 35, 41
arrow in flight paradox, 26–28
arrows of time, 24, 30
artificial intelligence, 139–148
astronauts, 19, 199

astronomers, 96
atom bombs, 83
atomic mass unit, 98
atoms, 2, 30, 39, 45, 57, 76, 97, 99
Autobiography of a Yogi
(Yogananda), 14
Avatar, 150
averages, law of, 90
awareness, 2, 94–95, 143, 144

B
Babylonians, 16–17
barium borate crystals, 67
baseball, 77
Baudrillard, Jean, 159
"Because I could not stop for
Death" (Dickinson), 171
Being, 167
Being, concept of, 26
Bell, John, 57, 71
Berkeley, George, 127
Berman, Bob, 109–112
beryllium ions, 64, 196
Bhagavad Gita, 13
Bible, 2, 10–11, 92, 108, 109

ABOUT THE AUTHORS

Robert Lanza, MD, is one of the most respected scientists in the world—a *U.S. News & World Report* cover story called him a "genius" and "renegade thinker," even likening him to Einstein. He is head of Astellas Global Regenerative Medicine, Ocata Chief Scientific Officer, and adjunct professor at Wake Forest University School of Medicine. Lanza was recognized by *Time* magazine in 2014 on its list of the "100 Most Influential People in the World." *Prospect* magazine named him one of the Top 50 "World Thinkers" in 2015. He is credited with several hundred publications and inventions and over thirty scientific books, including the definitive references in the field of stem cells and regenerative medicine. A former Fulbright Scholar, he studied with polio pioneer Jonas Salk and Nobel Laureates Gerald Edelman and Rodney Porter. He also worked closely (and coauthored a series of papers) with noted Harvard psychologist B. F. Skinner and heart transplant pioneer Christiaan Barnard. Dr. Lanza received his undergraduate and medical degrees from the University of Pennsylvania, where he was both a University Scholar and Benjamin Franklin Scholar. Lanza was part of the team that cloned the world's first human embryo, as well as the first to successfully generate stem cells from adults using somatic-cell nuclear transfer (therapeutic cloning). In 2001 he was also the first to clone an endangered species, and recently published the first-ever report of pluripotent stem cell use in humans.

Bob Berman is the longtime science editor of the *Old Farmer's Almanac,* and contributing editor of *Astronomy* magazine, formerly with *Discover* from 1989 to 2006. He produces and narrates the weekly *Strange Universe* segment on WAMC Northeast Public Radio, heard in eight states, and has been a guest on such TV shows as *Late Night with David Letterman.* He taught physics and astronomy at New York's Marymount College in the 1990s and is the author of eight popular books. His newest is *Zoom: How Everything Moves* (2014, Little Brown).

Don't miss the book that started it all, and shocked the world with its radical rethinking of the nature of reality.

BIOCENTRISM

"An extraordinary mind. . . . [Dr. Robert Lanza's] theory of biocentrism is consistent with the most ancient traditions of the world which say that consciousness conceives, governs, and becomes a physical world. It is the ground of our Being in which both subjective and objective reality come into existence."

—DEEPAK CHOPRA

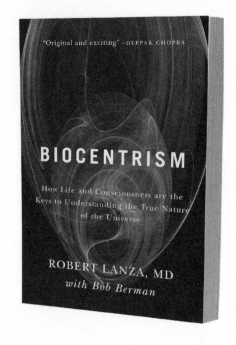

In *Biocentrism*, Robert Lanza and Bob Berman team up to turn the planet upside down with the revolutionary view that life creates the universe instead of the other way around.

Biocentrism takes the reader on a seemingly improbable but ultimately inescapable journey through a foreign universe—our own—from the viewpoints of an acclaimed biologist and a leading astronomer. It will shatter the reader's ideas of life—time and space, and even death. At the same time it will release us from the dull worldview of life being merely the activity of an admixture of carbon and a few other elements; it suggests the exhilarating possibility that life is fundamentally immortal. *Biocentrism* awakens in readers a new sense of possibility, and is full of so many shocking new perspectives that the reader will never see reality the same way again.

Learn more at
RobertLanzaBiocentrism.com